Engineering: A Very Short Introduction

VERY SHORT INTRODUCTIONS are for anyone wanting a stimulating and accessible way in to a new subject. They are written by experts and have been translated into more than 40 different languages. The series began in 1995 and now covers a wide variety of topics in every discipline. The VSI library contains nearly 400 volumes—a Very Short Introduction to everything from Indian philosophy to psychology and American history—and continues to grow in every subject area.

Very Short Introductions available now:

David Blockley

ENGINEERING

A Very Short Introduction

OXFORD
UNIVERSITY PRESS

Great Clarendon Street, Oxford, OX2 6DP,
United Kingdom

Oxford University Press is a department of the University of Oxford.
It furthers the University's objective of excellence in research, scholarship,
and education by publishing worldwide. Oxford is a registered trade mark of
Oxford University Press in the UK and in certain other countries

© David Blockley 2012

The moral rights of the author have been asserted

First Edition published in 2012

Impression: 6

Published in the United States of America by Oxford University Press
198 Madison Avenue, New York, NY 10016, United States of America

British Library Cataloguing in Publication Data

Data available

ISBN 978-0-19-957869-6

Printed in Great Britain by
Ashford Colour Press Ltd, Gosport, Hampshire

Contents

Preface

The purpose of this book is to explore, in non-technical language, what engineering is all about. The central message is that engineering is intimately part of who and what we are. We will explore the role of engineering in the modern world through the story of its history and relationships with art, craft, science, and technology. We will see how engineers utilize natural phenomena to embrace human needs. The story continues – so we look at some of the contributions engineers will have to make in the future in order to sustain and promote human well-being.

A recent UK parliamentary committee defined engineering as 'turning ideas into reality' – succinct but perhaps not that helpful. A dictionary definition is 'the art, science, and craft of changing a dream into a reality through conception, feasibility, design, manufacture or construction to operation and eventual decommissioning of something that fulfils a human need'. Engineering is the discipline of using scientific and technical knowledge to imagine, design, create, make, operate, maintain, and dismantle complex devices, machines, structures, systems, and processes that support human endeavour.

Such definitions are necessarily brief, and so they cannot convey the massive contribution engineering makes to modern life. Engineering is part of almost everything we do – from the water we drink and the food we eat, to the buildings we live in and the roads and railways we travel on; from the telephones and computers we use to communicate to the television, radio, music, and films that entertain us; from the bicycles and cars we ride to the ships we sail in and the aeroplanes we fly, from the gas and electricity we rely on to the power stations that generate the energy; from the X-ray machines that help doctors diagnose diseases to the heart pacemakers and artificial hips that help many people have a better quality of life. Unfortunately, it isn't all good. Engineering also contributes to the weapons we use to kill each other and the carbon dioxide emissions that threaten the planet.

The range of subdisciplines within engineering is large – there are over 30 different professional institutions in the UK alone that qualify engineers. Many, but not all, countries require state or provincial registration. Engineering activities vary from the 'one-off' massive infrastructure projects of civil engineering, including roads, bridges, reservoirs, and buildings, to the mass production of small mechanical and electronic components such as pumps, valves, pipes, motors, and integrated circuits and their assembly into larger manufactured products such as washing machines, cars, and aeroplanes. Over the last 200 years, these subdisciplines have tended to grow apart as scientific knowledge and technical achievement have seemed unstoppable. However, the complexity arising from the needs of the modern world has required a new coming together of specialists into teams that can tackle large projects in an integrated way. The modern engineer needs to work to achieve long-term sustainable development. We have become vulnerable to unexpected events, such as the ash cloud from an erupting volcano in Iceland that stopped all air traffic in northern Europe in April 2010 and the effects of the earthquake and consequent tsunami damage to a nuclear reactor in Japan in 2011. Coping with the massively complex interdependencies between systems is one of the major engineering challenges of the 21st century.

Engineering is a big subject and, like me, almost all engineers specialize to an extent. The choice of what and what not to include is entirely mine. I am sure that some will think I have not done justice to a particular subdiscipline, skirted over, or even missed out altogether, something essential. I have tried to be sufficiently representative to give the non-engineer reader a sense of scope, but I have also included some technical detail in order to give a sense of depth as well. Inevitably, I have left large gaps. I am fortunate to have been helped by many people. First and foremost, I would like to thank those who helped me make the book more accessible to a wider readership. Anne Thorpe read the whole book and made very many suggestions as to how it could be improved. Likewise, Joanna Allsop also read some of the text and made some suggestions. Angela Hickey explained to me some of the niceties of operatic voice control. Tim Cripps helped me with heart pacemakers. Simon Fieldhouse put me in touch with some developments in dentistry. Michael Liversidge helped me interpret some of the Italian artist engineers, such as Leonardo da Vinci, and Richard Buxton helped on some classical Greek words.

I would like to thank a number of engineers who were very generous of their time to help me to cover their specialist areas. Joe Quarini was an enthusiastic guide through thermodynamics and steam engines. He, Sandy Mitchell, and particularly Neil Brown helped me make some sense of jet engines. Mike Barton, Derrick Holliday, and Duncan Grant were very friendly, willing advisers on electromagnetism. Mike and Derrick read drafts and were particularly patient when I peppered them with many questions. Figure 15 was arranged by Derrick and photographed by Richard Walker, and Figure 17 was provided by Mike. Joe McGeehan helped me with wireless communication and David Stoten with control engineering. Patrick Godfrey read and commented on Chapter 6 – systems thinking. Bob Baird and Mike Rogers read the whole book to help me incorporate some chemical engineering and to give a North American perspective. Mike provided Figure 20. Robert Gregory also read the whole text, made several suggestions, and supplied Figure 9. Figure 1 is reproduced by kind permission of Trevor Harrison, warden of Selsley Church, UK. Figure 2 is an Evia pacemaker and is printed with the permission of Biotronik, a leading global company in the field of biomedical technology with a focus on electrotherapy of the heart and vascular intervention. Figure 6 is reproduced by kind permission of the Technical Director of Caudwell's Mill, Rowsley, Derbyshire. Finally, I would like to thank Rolls-Royce plc for permission to use Figures 10 and 11. The copyright of these images is owned exclusively by Rolls-Royce plc.

Finally thanks are due to Latha Menon at Oxford University Press who had the idea for this book, Emma Marchant, Elmandi du Toit, and Carol Carnegie also at OUP, Subramaniam Vengatakrishnan at SPi Global, copy-editor Alyson Silverwood and proofreader Kay Clement. Last, but by no means least thanks to my wife Karen for her unfailing love and support and endless cups of tea.

List of illustrations

Chapter 1
From idea to reality

Engineering is, in its most general sense, turning an idea into a reality – creating and using tools to accomplish a task or fulfil a purpose. The word 'engineer' derives not, as you might imagine, from being-someone-who-deals-with-engines but rather from its Latin root *ingeniarius*, Old French *engignier*, and Middle English *engyneour* to mean someone who is ingenious in solving practical problems. Man's ability to make tools is remarkable. But it is his ingenious ability to make sense of the world and use his tools to make even more sense and even more ingenious tools, that makes him exceptional. To paraphrase Winston Churchill, 'we shape our tools and thereafter they shape us'. Tools are part of what it is to be human. In the words of Henry Petroski, 'To engineer is human'.

Yet for many, there is a disturbing cloud. Once tools were simple common sense – almost all were understandable to the intelligent layperson. Since the Industrial Revolution, the interior workings of many tools have become mysterious, complex, and opaque to all but specialists. The culture of opposition between the arts, religion, science, and technology has widened and is often antagonistic. Matthew Crawford has accused engineers of hiding the works, 'rendering many of the devices we depend on every day unintelligible to direct inspection'. On the other hand, Brian Arthur has described a process he calls 'structural deepening' in which engineers inevitably add complexity as they strive to enhance performance. There is an increasing realization that science, technology, engineering, and mathematics (STEM) are intertwined in a way that needs disentangling. This is urgent because engineering is so intimately part of who we are that effective democratic regulation requires us to understand something of what engineering offers, what it might offer in the future, and, perhaps most importantly, what it cannot ever deliver.

Our journey through engineering will be broad and deep. We will explore in this chapter the relevance of engineering to many aspects of modern life, including fine art and religious faith. Depth will be manifest in later chapters, as we investigate how complex physical processes have been harnessed, for example to make a transistor. We will discover the excitement of facing difficult and important challenges, such as the spanning of a large river with a big bridge. We will become more contemplative in the final chapter, as we probe the creative thinking needed to engage with the new age of systems complexity and the need for a fresh approach to dealing with uncertainty.

All through history, people have expressed their awe, wonder, spirituality, and faith by making tools – especially for religious ritual. It was one way of coping with fear, pain, and the mysteries of the unknown. Pyramids connected this world to the afterlife. Church art, paintings, and sculpture were ways of creating a mood for devotion and the telling of the Christian story. From ancient tablets of stone to present-day LCD (liquid crystal display) screens, from horse-drawn carts to space rockets, from stone arches to cable-stayed bridges, from pyramids to skyscrapers, from the moveable type of the printing press to the integrated circuits of computers, from the lyre to the music synthesizer, from carrier pigeons to the internet, our tools have become ever more powerful and conspicuous.

Imagine waking up in the morning, attempting to switch on your light or radio, but finding that everything provided by engineers had gone. Many disaster movies rely on this kind of idea. Suddenly there is no electricity, no heating, and no water through the tap. You have no car, and there are no buses, trains, no buildings and no bridges. At the beginning of the 21st century, almost everything we rely on is a product of the activities of engineers and scientists.

Of course, just because you rely on something doesn't mean you must find it interesting. After all, few drivers want to know in detail what is happening under the bonnet of a car. Matthew Crawford, in his book *The Case for Working with Your Hands*, argues that each of us is struggling for some measure of self-reliance or individual agency in a world where thinking and doing have been systematically separated. We want to feel that our world is intelligible so we can be responsible for it. We feel alienated by impersonal, obscure forces. Some people respond by growing their own food, some by taking up various forms of manual craftwork. Crawford wants us to reassess what sort of work is worthy of being honoured, since productive labour is the foundation of all prosperity. Of course, technical developments will continue, but as Crawford argues, we need to reassess our relationship with them. Just as a healthy lifestyle is easier if you have some empathy with your body and how it is performing, so you might drive better if you have some rapport with the workings of your car. In the same way, perhaps a fulfilled life is more likely if you have some harmony with the things you rely on and some feeling of why sometimes they don't perform as you might wish.

One only has to think of the railways, the internal combustion engine, the contraceptive pill, the telephone, the digital computer, and social networking to realize that technical change profoundly affects social change. Economists such as Will Hutton and Richard Lipsey argue that technological development is fundamental to economic progress. Historians argue about the drivers of the British Industrial Revolution of the late 17th and early 18th centuries, but all recognize the profound social changes resulting from the exploitation of coal and iron, the building of canals, the development of clever new machines such as the spinning jenny and cotton gin for the textile industry. Nevertheless, many people still regard the products of engineering as morally and ethically neutral – in other words, they are intrinsically neither bad nor good. What is important, according to this view, is how we humans use them. But engineering is value-laden social activity – our tools have evolved with us and are totally embedded in their historical, social, and cultural context. Our way of life and the objects we use go hand in hand – they are interdependent parts of our culture. Transport is a good example. The canal network opened up possibilities for trade. The steam engine and the railways created new opportunities for people to travel. Communications were transformed. The consequent social changes were large. They affected the places people chose to live. They enabled people to take seaside holidays. They transported natural resources such as coal and iron much faster than canal barges. Different kinds of fresh food became widely available. Newspapers and mail were distributed quickly. These social changes led to a cultural 'climate' where further technologies such as the internal combustion engine and road transport could flourish.

Whilst it may seem impossible for non-specialists to influence technological developments of this kind, informed debate about them is essential in a healthy democracy. A modern example is our collective response to the threat of climate change. Regardless of the rights and wrongs of the arguments about whether we humans are contributing to a global average temperature rise, there seems to be little doubt that we are going to experience a period of more and more extreme weather events. Climate change does not threaten the planet – it threatens us and our way of life. If we do not cope well, millions of people will suffer unnecessarily. We have collectively to understand better what we can and cannot expect of our infrastructure in the future so we can manage the risks as effectively as we can. That requires an understanding of what it is reasonable or unreasonable to expect of engineering and technology.

Engineering directly influences the way in which we humans express our deepest emotions in religion and in art. The London Millennium Bridge is just one example of building as a way of commemorating a significant anniversary – it isn't just a river crossing. Even when we want to express naked power, we build structures – the old medieval castles with drawbridges are examples. Modern skyscrapers demonstrate the economic power of multinational companies. A completely different example is how technology has changed musical styles. Amplification and projection of the voice of an opera singer derives from resonance in the head, the chest, and indeed the whole body. *Bel canto*, or the art of beautiful singing, in 17th- and 18th-century Italy cultivated embellishment through ornaments such as cadenzas, scales, and trills which declined as the size, power, and volume of the orchestra increased. By the end of the 19th century, singers had to find even more power, but then came recording and electrical amplification, whereby balance between singer and orchestra could be achieved by twiddling a knob. Now singers of popular music use different techniques of larynx position and abdominal control of breath because sheer power is not needed. Much of modern music uses amplified electronic equipment, enabling new ways of self-expression.

If engineering is about making and using tools, then we need to be very clear what we mean. A tool is anything used to do work, and work is effort or exertion to fulfil a purpose. A hammer, a drill, and a saw are obvious examples. Tools in an industrial workshop may be a lathe, a welding torch, or a press. In each case, the work being done is clear. Less evidently perhaps, other tools in the home such as a kettle, a cooker, and a refrigerator do work by heating and cooling. We will explore work and heat in Chapter 3 when we look at engines that use heat to power planes, boats, trains, and cars. Back in the home, we can even regard chairs as tools for sitting, although it is not obvious how chairs do any work. In fact, as we will explore in Chapter 2, they do it by responding to your weight as you sit down with 'ever-so-slight' movement.

Bridges and buildings respond to the traffic passing over them in a very similar way. Buildings are tools for living. Obvious office tools include a desk, pencil, and paper, but a modern office requires telephones, computers, and the internet. We use these tools to process information and the work they do is electromagnetic, as we shall see in Chapters 4 and 5. Electronic musical instruments, media, arts, and entertainment such as sound and video recordings, radio, television, and mobile communications all rely on electromagnetic work.

Some tools are very obvious, like a screwdriver, and some not so obvious, like an artificial hip. Some are big, like a bridge or a water reservoir, whereas some are very small, like a silicon chip or a wood screw. Some are useful, like a food mixer, and some are mainly for entertainment, like a television or DVD player. Some are destructive and involve significant ethical issues, like weapons of war or torture. Some tools are complex systems, like an airport or the internet, and some are simple, like a safety pin or a paper clip. However, they all have one thing in common – they work to fulfil a human purpose, whether good or malevolent.

It is indisputable that engineering is at the heart of society. But perhaps you are already thinking that all of these things – especially computers and the internet – are the work of scientists. As we said earlier, the words science, technology, engineering, and mathematics are often used interchangeably even though good dictionaries are clear enough. They tell us that science is a branch of knowledge which is systematic, testable, and objective. Technology is the application of science for practical purposes. Engineering is the art and science of making things such as engines, bridges, buildings, cars, trains, ships, aeroplanes, chemical plant, mobile phones, and computers. Mathematics is the logical systematic study of relationships between numbers, shapes, and processes expressed symbolically. Art and craft are closely related. Art is difficult to define but is a power of the practical intellect, the ability to make something of more than ordinary significance. Craft is an art, trade, or occupation requiring special skills – especially manual skill. But the boundaries are not obvious – for example, was Stradivari a craftsman or an artist? The outstanding qualities of his violins have yet to be surpassed by modern techniques. So put simply, science is what we know, art is making extraordinary things, engineering is making useful things, technology is applied science, mathematics is a tool and a language, and craft is a special skill. It seems, therefore, that we have to conclude that the cloud around these terms derives from the history of their development.

Our achievements based on science speak for themselves. Cool, clear rational thinking works. But does it work for everything? Science is powerful because we can coordinate its various parts so that ideas fit together into a coherent whole. We can share it – it isn't subjective and personal like an emotion. We can repeat it – it isn't a single experience but is something that we can test and validate in independent ways. However, science is not complete – it is not an all-powerful way of getting at the absolute Truth about the actual physical world 'out there'. Our understanding depends on context in subtle and complex ways. It can always be, and is being constantly, improved. Whenever we act on scientific knowledge, there is inevitably the possibility of unintended consequences, as the philosopher Karl Popper pointed out.

Then there are whole areas of our lives where science has nothing to offer – it cannot assuage pain or sorrow in times of personal loss – it has no sense of tragedy and no sense of humour. Art, craft, and religion can help to give life meaning and purpose. Many of us use and develop our craft skills to draw, photograph, cook, or garden – the satisfaction is private through engaging with the material world to do something well. Arts and crafts have a strong connection with technology. For example, at the turn of the 19th century, the arts and crafts movement recognized the inventiveness and imagination in craft and technology but saw the Industrial Revolution as a threat to artistic creativity and individuality.

1. The South Window at Selsley Church, UK, by William Morris 1861–2

For John Ruskin, a healthy society, both morally and socially, depended on skilled and creative workers. The arts were to be judged according to the amount of freedom of expression they allowed to the workmen; perfection and precision were suspect since they implied direction and repression. It was a socialist idea especially promoted by William Morris. He designed and sold products such as stained-glass windows, wallpaper, and ceramic tiles (Figure 1). In his lectures, Morris said:

> Nothing can be a work of art that is not useful; that is to say, which does not minister to the body when well under the command of the mind, or which does not amuse, soothe, or elevate the mind in a healthy state.

> That thing which I understand by real art is the expression of man of his pleasure in labour....As to the bricklayer, the mason, and the like – these would be artists, and doing not only necessary but

beautiful, and therefore happy, work, if art were anything like what
it should be.

But arts and crafts aren't static – they develop as new technologies
create new freedoms. Just as painters like J. M. W. Turner were
able to take advantage of new pigments, so 21st-century artists are
now using digital electronics.

In everyday use, the word 'engineering' has many meanings. For
example, we use it in the phrase to 'engineer' a solution to a
problem. People 'contrive to bring about something', or 'skilfully
originate something', presumably to address a need or purpose.
So engineering involves creative problem-solving. We engineer
an agreement, a deal, a plan, or even a 'bright future' for
ourselves.

In the 'olden' days, it was common for young boys to want to grow
up to be engine drivers. The reopening of numerous steam
railways around the world by enthusiasts demonstrates the
romantic fascination many people still have for the age of steam.
Some reputable dictionaries still define an engineer as 'a person
trained and skilled in the design, construction, and use of engines
or machines' or 'a person who operates or is in charge of an
engine'. So another meaning of engineering is 'looking after
engines and machines'.

If we have a problem with a machine or piece of technical
equipment such as a gas boiler, a washing machine, or a
photocopier, then in the UK we send for the engineer to come and
fix it. If there is a power cut because extreme weather has brought
down the power lines, then the media will tell us that the
engineers are fixing the problem. So yet another meaning of
engineering is 'fixing a problem'.

When an industrial dispute, say in construction engineering, car
production, or other manufacturing, deteriorates into industrial

action, the banner newspaper headlines might tell us 'engineering workers out on strike'. So engineering is also about 'working in industry'.

When a new bridge is opened, we might be told that the bridge was designed by an architect and by an engineer. We might be told that a new car is well engineered. Engineers design aeroplanes and computers. Yet designing, making, and selling new technical products from bridges, cars, mobile or cellular phones to power stations involves very many people – many of whom are not engineers. So engineering is being part of a team 'providing significant practical complex things'.

Complex things like computers cannot be designed without using the latest knowledge – the appliance of science – technology. If you look at the technical papers reporting this work in professional and research journals, you will find them very difficult to follow if you are not technically qualified – they are full of abstruse words and mathematical equations that even technical experts in parallel subjects can find difficult. So engineering is 'applying science' – usually in a very narrow field of application. Indeed, the field may be so narrow that engineers are sometimes accused of being technically narrow and narrowly technical – even to the extent of being 'nerds' or 'anoraks' – intelligent but single-minded and obsessed.

So there are at least these six interpretations of the word 'engineering' – but there are more. Just look at some of the job titles of engineers in Box 1. Of course, engineers have to specialize – it is pointless calling a photocopier engineer to fix a gas boiler. But when we list some of them, we find a bewildering array of types. Box 1 lists only some of the titles you can find. There seems to be an enormous variety. Can we classify them in any sensible way? Note that some of the types are in bold and some in italics – we will come back to why later.

Box 1 Some of the various types of engineer

Acoustic, aeronautical, aerospace, agricultural, asphalt, automotive, biomechanical, biomedical, bridge, building, cast metal, *chartered*, **chemical, civil, computer** (software, hardware), concrete, construction, cost, control, corrosion, craft, dam, design, diesel, dynamics, **electrical**, electronic, *engineering worker*, environmental, explosives, finance, fire, gas, harbour, healthcare, heating, highway, *incorporated*, industrial, information, instrumentation, knowledge, lighting, marine, material, measurement and control, **mechanical, medical**, mineral, mining, motor, municipal, naval architects, non-destructive testing, nuclear, operations, photonics, plumbing, power, production, *professional*, project management, quality, railway, refrigeration, *registered*, reliability, river, robotic, royal (military), safety, sanitary, sensor, signal processing, space, structural, sustainability, systems, *technician*, transportation, turbine, welding, water.

So to recap, we know that some engineers look after machines and engines, some fix problems, some work in industry, some are responsible for significant practical complex things, some are applied scientists, and some just make something happen to address a need. But what is it that is common? How do we go about classifying engineering activity?

One way is to distinguish the different scope and responsibility of work from the specific industrial expertise required. The scope of work ranges from creating the first idea to making the final reality – from the conceptual dreaming of the visionary designer to the harsh practical manual labour required to put the idea into practice – from the idea of a building to the tasks of laying the bricks and wiring up the electrics. In the UK, four levels of scope are described as *engineering worker*, *technician engineer*, *incorporated engineer*, and *chartered engineer* – listed in italics in Box 1. Unfortunately, different names are used in different countries, but it is common to describe them all by the one word 'engineer'. In the USA, the term 'chartered engineer' is not used – the (not exact) equivalents used there and in many parts of the world are *professional* and *registered engineer*. These levels are important protections for public safety and in most (but not all) countries are regulated by law. They are used to define jobs, qualifications, and career paths to make sure only those suitably qualified can take on particular responsibilities.

The six horizontal divisions or categories of engineering activity are civil, mechanical, electrical, chemical, computing, and, more recently, medical engineering. The names are reasonably self-explanatory – except for civil. In the 18th century, all non-military engineering was called civil engineering. However, as the need for more specialist engineers developed, so civil engineering has come to refer to construction and infrastructure engineering. One example of how these various types of engineers work together is providing us with clean drinking water.

Civil engineers build and maintain reservoirs, including the dams that control river flow, to store water. They install the large pipes that distribute the water to various locations such as the treatment works. Here chemical engineers and scientists supervise the detailed processes of filtering out large objects (like logs of wood or dead animals) and making controlled use of biological (microbes that decompose organic matter) and chemical processes to clean the water and make it potable. They also constantly monitor the quality of the water. Water flows under gravity but sometimes has to be pumped. Pumps and many other pieces of mechanical plant, machinery, and equipment such as valves, flow meters, water meters, and filtration plant are supplied, installed, and maintained by mechanical engineers. But much of this equipment is powered by electricity, so without electrical engineers we would have no water. Electronic engineers provide the instrumentation used to monitor and control the flow of water. Computer systems are designed, installed, and maintained by computer engineers to control the flow of information necessary to keep the water flowing.

Water is basic to human life and not just for drinking. In hospitals, clean water is essential for washing and cleaning surgical equipment and wounds. Hospital buildings and equipment have long depended on engineering, but the more recent direct collaborations between doctors and engineers have produced exciting and important life-improving and life-saving developments. An example is the treatment of osteoarthritis – a common problem for many people after middle age. It is a degenerative arthritis due to wear and tear, commonly in the hip joint. Surgeons can use a miniature TV inserted into a joint though a small incision to check the condition of the cartilage, and at the same time they can try to relieve the pain by cleaning or flushing out the joint. Artificial hip replacements can reduce pain, and improve movement and quality of life. New joint surfaces are created between the upper end of the thighbone using a metal ball and the hip socket in the pelvic bone with a metal shell and plastic liner. The joints may be glued to the existing bone. Alternatively, a porous coating that is designed to allow the bone to adhere to the artificial joint is used such that, over time, new bone grows and fills up the openings.

In 2009, it was reported that about half a million heart pacemakers (Figure 2) have been inserted into patients in the UK alone since the first one in 1958.

Normally, your heartbeat is controlled by your heart's own bioelectrical triggering system. If this stops working properly, then a pacemaker can help. The most commonly installed pacing device is a demand pacemaker. It monitors the heart's activity and takes control only when the heart rate falls below a programmed minimum – usually 60 beats per minute. The first devices

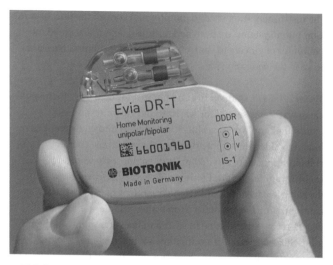

2. Evia heart pacemaker

had to be worn around the neck with wires connecting to the heart – some were even plugged into the mains electricity. Although these early examples were large, now they are typically the size of a match box and inserted under the skin. The pacemaker has two parts – a battery-powered generator and the wires that connect it to the heart. The generator is implanted just beneath the skin below the collarbone. The leads are threaded into position through veins leading back to the heart. They are programmed from a small computer and their batteries last more than seven years without a recharge. The entire implantation procedure requires only a local anaesthetic and takes about an hour.

Although these six classifications of the scope and responsibility and specific engineering expertise are interesting and useful, they come from within engineering itself and they don't help us to disentangle STEM. So if we cannot find clarification from within engineering, then we must look for it from scholars outside of engineering – philosophers and social scientists, for example. After all, the philosophy of science is a well-developed and major intellectual discipline. Unfortunately, apart from a few exceptions, scholars say little about engineering. Of those who do, most seem to assume that engineering is technology and that technology is applied science.

A major exception is Carl Mitcham, who has examined technology from four perspectives – as objects, as knowledge, as activity, and as an expression of human will. The first three are reasonably obvious. Clearly, technical objects are artefacts – the engineering tools we listed earlier – not forgetting to include artworks and religious works such as paintings, sculpture, and musical instruments. Engineering knowledge is specialized and it works. Indeed, the success of technology is often quoted by philosophers as evidence of the truth of science. But, as we shall see in later chapters, the story of the relationship between what we know and what we do is not so straightforward. Technological activity includes crafting, inventing, researching, designing, making, operating, maintaining, and decommissioning. Mitcham's final perspective, human will or volition, is less obvious perhaps. But here lies the key to understanding the central differences within STEM. Needing, willing, wanting, desiring, or wishing defines purpose. As our story unfolds in the next chapters, we shall see that whilst methods may seem indistinguishable, individual purposes within STEM may be quite distinct.

The many ways in which we learn to 'know' and we learn to 'do' evolve in leapfrog fashion. Babies 'act' before they 'know' – though clearly we are born with inherited genetic innate skills and knowledge by which we learn quickly to grow and develop. But 'doing' comes first in the sense that we act before we become aware. As our brains develop, we learn to speak and think and become self-conscious beings. Through that learning, we behave differently and through that behaving we learn to think differently.

Living is primarily about 'doing', based on our developing knowing or learning. In that sense, engineering is a form of living since 'doing' or practice is prior to knowing, but knowing informs better practice. So we can think of living as a form of engineering with the end product of human flourishing within an engineered political system.

Confusion about the words 'engineering' and 'technology' often derives from different uses by different people from different backgrounds in different contexts. We therefore have to be very careful to make clear what we mean. In this book, I will use engineering and technology as synonyms but distinct from science and mathematics in their central purpose. So in summary, the purpose of science is to know by producing 'objects' of theory or 'knowledge'. The purpose of mathematics is clear, unambiguous, and precise reasoning. The purpose of engineering and technology is to produce 'objects' that are useful physical tools with other qualities such as being safe, affordable, and sustainable. All are activities arising from human will that sustains our sense of purpose. Science is an activity of 'knowing', whereas engineering and technology are activities of 'doing' – but both rely on mathematics as a language and a tool. The methods they adopt to achieve their purposes are so very similar that, unless you understand their motive and purpose, it is often unclear whether a given person is behaving as a scientist or as an engineer/technologist.

The story of engineering naturally divides into five ages – gravity, heat, electromagnetism, information, and systems. The first three are the natural phenomena that scientists study and that engineers and technologists use to make their tools. From the ancient skills used to build pyramids from natural materials to the modern engineering of skyscrapers, we have systematically developed our scientific understanding of gravity and used it to build bigger, higher, and longer. Our primitive control of fire has developed into mechanical and chemical power from heat through steam, internal combustion, and jet engines and manufactured materials. Electromagnetism is a relative latecomer in the long history of human development which has given us electricity, motors, computers, and telecommunications. Out of this came the age of information, which has turned now into the age of complex systems. In the last chapter, we will see how 'systems thinking' is helping us to integrate disparate specialisms by seeing tools as physical 'manipulators' of energy embedded in 'soft' people systems. From the science of Aristotle to Newton to Einstein, from the craft of Vitruvius to Leonardo to William Morris, and from the engineering of Archimedes to Faraday, to Berners-Lee and the World Wide Web, the story of engineering is racing ahead at an ever-increasing pace. It is a story that has had, and is still having, a profound influence on the quality of human life.

Chapter 2
The age of gravity – time for work

What have a child's swing, a golf club, an opera singer, a flute, a radio, and a bridge got in common? The answer is timing – getting the best out of something with the least input of work. Imagine pushing the child's swing – you quickly learn timing, i.e. when to shove for maximum effect. In effect, you tune yourself to push the swing at what scientists call its natural frequency, i.e. the number of times it goes back and forth in an amount of time (usually minutes or seconds) when swinging freely. Your pushes are (an external stimulus) timed to produce resonance (a large amount of swing or amplitude of vibration) because you are pushing at a frequency close to the natural frequency of the swing.

A golf club has a natural frequency too. If a golfer, teeing up for a shot, can match his swing with the natural frequency of his club as it flexes in his hands, then he will have a 'sweet' shot – something all amateur golfers aspire to and talk about when it happens. He will use the elasticity of the club to transfer the energy of his back swing into the ball with maximum effect. The same phenomenon occurs when a baseball player or a cricket batsman finds the sweet spot on the bat. You will recall in Chapter 1 that we said that an opera singer can project his or her voice even in a large concert hall whereas a 'pop' singer has to have a microphone. Again, it's all about timing. A trained tenor can make the sound of his voice resonate through his head and chest cavities by altering the shape of his mouth, tongue, and lips, his breathing, and the movement of his larynx. When a flute player blows air into the mouthpiece of a flute, the air in the body of the flute resonates and produces a musical note – timing again. Different notes at different natural frequencies are made by fingering over the holes in different combinations. Stringed instruments rely on the resonance of vibrating strings of different length and mass. As we shall see later in Chapter 4, when you tune a radio to a particular radio station at a particular frequency, you are actually altering the electrical impedance (opposition to the electrical current) of an electrical circuit. You do this until the circuit resonates with (is in time with) the frequency of the electromagnetic radiation being received by your aerial from your chosen station.

These are all useful examples of resonance – but sometimes resonance is a dangerous nuisance and the best timing creates the least effect. Bridge designers must design their bridges so that they do not resonate when soldiers walk in step or the wind blows because the large vibrations can cause excessive damage. The Angers Suspension Bridge in France collapsed in 1850 as 500 soldiers were marching across and 220 were killed. We want a bridge and the wind blowing onto it not to be synchronous, i.e. out of timing. Perhaps the most famous commonly quoted example was the collapse of the Tacoma Narrows Bridge in 1940. However, modern wind engineers attribute that collapse to a different but related phenomenon, called self-excited flutter from the shedding of eddies or vortices.

So timing is important in the way most tools do work. Our own idea of work is familiar – exertion, effort, labour, or toil to make a living. It is every effort we make ranging from just moving around – like getting out of bed – to hard physical exercise such as running 10 kilometres. We work when we lift a heavy weight like a shopping bag. We work when we think hard about a problem. In more abstract terms, we work whenever we make a change of any kind. It is easy to see that when we push a child's swing, sing opera, or play a flute, we are doing work – but the kind of work being done in a radio or a bridge is not so obvious. In this and the next chapter, we'll see how a bridge does work against gravity as it carries traffic over a river, and in Chapter 5 we'll examine electromagnetic work in a radio.

In STEM (science, technology, engineering, and mathematics), work is defined precisely and objectively so that it is unbiased and independent of personal opinions. Work is the product of force and distance. So if you lift a mass of 1 kilogram (which has a weight of approximately 10 newtons) through 2 metres, then the work you do is 20 newton metres, or 20 joules. In the past, before the widespread adoption of SI units, the work that engines were capable of doing was compared with the work that horses could do – hence the term 'horsepower'. Various people came up with various equivalences, but the modern agreed definition is that 1 horsepower is 746 joules per second or 746 watts. When we feel energetic, we feel ready to work – so energy is the capacity to do work and is also measured in joules. Power is the rate of expending energy or doing work. It is measured in joules per second or watts.

We will start our examination of engineering in depth by looking at some of the earliest tools that have to work against the forces of gravity, and as we do we will see how the special role of the engineer evolved. Figure 3 is a timeline to help you keep track of the order in which things happened. In Chapter 3, we will look at modern examples together with how we use heat to do work and work to make heat, and in Chapters 4 and 5 we will explore how we use electromagnetic work to process information.

It is impossible to know just how the first humans conceived force and time. Clearly, they would have a common sense notion of work and effort, and they would have noticed the regular movements of the Sun, Moon, and stars. The weather patterns,

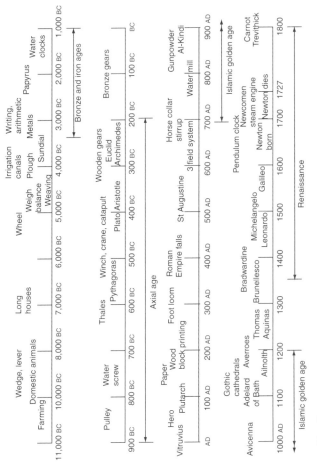

3. Timeline

including thunder and lightning, must have seemed beyond humankind, sometimes welcoming (warm sunshine) and other times threatening (thunder and lightning). So they began to make up stories about natural events. Their *need* to feel safe from these other-worldly events drove an *activity* of creating stories which were the *objects* that served as explanations or *knowledge*. Such stories were *mythos*.

These regularities were practically useful. The nomadic hunter-gatherers *needed* to find food and would have modified their hunting *activity* as they saw how the behaviour of animals and plants was related to the movements of the Sun in the daily cycle, the Moon in the lunar cycle, and the seasons in the annual cycle. At first, they probably relied on judging the position of the Sun in the sky or shadows on the ground. By sharing that *knowledge* and cooperating in the hunt, they learned to be more effective at tracking animals and finding water. They learned to control fire and, as caves couldn't be moved, they began to make tools, clothes, build huts of tree branches, leaves, straw, stones, and animal skins – *objects*.

Perhaps the first farmers (settled farming dates from around 10000 BC – domestication of animals around 8500 BC) used the length of shadows or a stick in the ground (gnomon) as a primitive sundial. As they began to barter their goods, they needed to know when to sow, when to harvest, when the rivers may flood. They needed to estimate the size of their fields and their crops. By 7000 BC, large buildings, or 'longhouses', up to 30 metres long were being built across northern Europe.

During the 4th and 3rd millennia BC, metals were obtained from ores, melted, cast, and hammered. Textiles began to be woven from flax and wool. Writing came about 3500 BC, as did the first recorded sundial found in Mesopotamia (now south-eastern Iraq). Arithmetic was being developed by 3000 BC in Egypt. The Sumerians who lived between the Rivers Tigris and Euphrates

4. Weighing scales, Luxor Temple, Egypt, 5000 BC

began to irrigate their fields with canals and ditches. In order to set these out, they needed to understand the likely water flows from these rivers to their crops.

To help all of this activity, five basic types of machines were developed in antiquity – the wedge, the lever, the wheel (including the winch and the gear or toothed wheel), the pulley, and the screw. The wedge was perhaps derived from the axe – one of the first Stone-Age tools. It was used for splitting wood and cutting stone slabs from quarries. Levers date back to prehistory and were used then, as now, for moving large objects and, for example, as hoes for cultivation, spades for excavation, and oars for rowing. Around 5000 BC, the lever, as a simple balance, was used for weighing (Figure 4). The date of the first wheel is unknown, perhaps also around 5000 BC. The winch or capstan is a wheeled drum or shaft that can be turned, by hand, using radiating spokes or handles. A heavy load is pulled by attaching it to a rope or chain wound around the drum. The ease with which a handle can be turned, relative to the heavy load being pulled, is called the mechanical advantage. In other words, it is the ratio of the output force (the heavy load) to the input force (the force needed to turn the handles) and is equal to the ratio of the radius of the spoke handle to the radius of the drum. Examples of cranes, catapults, and tread mills based on the winch date back to the 5th century BC, and improvements in the mechanical advantage of the machines were sought intuitively.

The inadequacies of sundials would have become obvious – they don't work at night or in the shade. So by around 1400 BC, water clocks were being used. Water was drained slowly from one container to another and the water levels shown by marks on the containers used to indicate the time. A gear or toothed wheel, made in wood, is first mentioned in Egypt around 300 BC and was used as one of the ways to improve water clocks. A float was connected to a notched stick which through a gear tooth wheel turned a hand that pointed to the time.

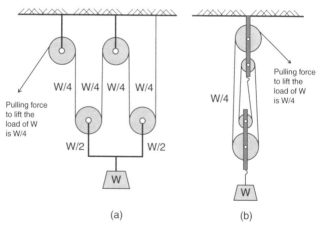

5. **Using pulleys to lift a weight**

The idea of a pulley was perhaps inspired by throwing a single rope over a tree branch. A single pulley, used in ships, water wells, and the like, is shown in an Assyrian relief from 870 BC. The compound pulley (Figure 5) is often attributed to Archimedes and was given detailed treatment by Vitruvius. A screw inside a pipe used for raising water may have been part of a pump for Sennacherib, King of Assyria, for the water systems at the Hanging Gardens of Babylon and Nineveh in the 7th century BC. However, it is more commonly ascribed to Archimedes (Figure 6) but probably invented by the Pythagorean Archytas of Tarentum. Wooden screws were commonly being used by the 1st century BC in, for example, oil and wine presses. A shipwreck found near the island of Antikythera in 1900 revealed a complex analogue device of more than 30 finely tuned bronze gear wheels for calculating time and astronomical cycles, built around 100 BC, probably in Rhodes.

ARCHIMEDEAN SCREW
The Archimedean Screw was the means used to transport
...t and grain horizontally from one side of the mill to
...her.

6. An Archimedean screw used to move grain in a flour mill, Derbyshire, UK

The period from 900 BC to 200 BC has been called the 'Axial Age' because it was pivotal to human development, not just spiritually but intellectually and practically. During this time, in four distinct regions, the world's great traditions came into being: Confucianism and Daoism in China, Hinduism and Buddhism in India, monotheism in Israel, and philosophical rationalism in Greece. The frontiers of human consciousness were pushed forward. It became essential to test everything empirically against personal experience. Religion was a practical matter – it was about how you behaved not what you believed. Technological developments were not simply the result of guesswork and luck. People showed a highly developed ability to observe and to learn from experience. They may have expressed their deepest thoughts in terms of *mythos* but their intellectual awakening enabled them to progress as *need* drove *activity*.

The ancient Greeks were deep thinkers and polytheists. From around the 8th century BC, they began to tackle the higher *needs* of esteem and self-fulfilment. One of their first great thinkers was Thales, born in the 7th century BC. He was a man of many talents – statesman, engineer, businessman, philosopher, mathematician, and astronomer. He suggested that the way to live a righteous life is to refrain from doing what we blame in others. He was practical – he helped an army cross a river by diverting the stream. He went to Egypt and brought some of their 'geometrical facts' back to Greece. He estimated the height of a pyramid by observing the length of the shadow of a pyramid at the same time as his shadow was the same length as his height. The Egyptians had rules for calculating areas of fields and volumes of crops and so on, but they had no concept of geometry as a systematic way of seeing relationships. Geometry was a Greek invention. The seeds for Western science were being sown. Bertrand Russell said that Western philosophy began with Thales. Geoffrey Lloyd said that there was a discovery of nature and the practice of rational criticism and debate. Traditional explanations that story-tellers had passed on without any real criticism now were in competition as people looked for the best explanations.

Where earlier thinkers had attributed material substances as primary things, about 50 years after Thales, Pythagoras established an esoteric sect (*activity*) who thought the patterns (*objects*) they discovered in numbers were the route to the divine (*knowledge*). For example, they discovered that musical harmonies depend on numerical ratios in the length of a string. The Pythagoreans thought that by concentrating on pure abstractions, they could 'wean themselves away from the contaminations of the physical world and get a glimpse of divine order'.

The Sophists, such as Hippias of Elis, emphasized common sense and *techne*, a technology that would make them more effective here and now. They wanted ordinary people to take full advantage of their own potential – to forget the old fairy tales and think for themselves. Detailed and methodological observations were valued. They sought to distinguish between people with medical knowledge and magicians, quacks, and charlatans. Greek surgeons set broken limbs using a winch.

Socrates had a mission to bring his fellow Athenians to a better understanding of themselves. Almost all we know of Socrates is through the writing of Plato (around 400 BC). Plato picked up on the Pythagorean focus on abstraction through his 'theory of forms'. A form is an archetypal essence of something beyond any actual manifestation of a reality – an abstract *object*. A circle is a good example since its definition is abstract and perfect, but every circle actually produced is inevitably imperfect – even if only very slightly. Over the door of Plato's academy was the motto 'Let no one unacquainted with geometry enter here'. The ideals of mathematical form were divine. The real world was untidy – only the world of forms was perfect (*knowledge*). It was a dimension of reality that transcended normal experience but was entirely natural. For example, Plato's description of beauty was, as Karen Armstrong notes, similar to what others called God or the Way – absolute, unique, eternal – but beauty was only part of the Good. But Plato's aim was not religious – he wanted a rational cosmology. It was a powerful vision that when later merged with monotheistic religion would influence Western thought profoundly.

Aristotle was Plato's most brilliant pupil, who brought philosophy down to earth. Instead of seeking meaning in the immaterial world, he found it in 'change'. Change was a universal striving for fulfilment. He explained time in terms of change and not vice versa. Human well-being lay in intelligent, clear, rational thinking – this was the way man linked with the gods and grasped ultimate truth. It was *logos*.

Aristotle maintained that everything is moved by something else – so there must be an unmoved mover – God. Reason demanded that a chain of cause and effect must start somewhere. Aristotle believed that the natural state of an object is at rest and that it won't move unless acted on by a force. The earth is at rest and so a heavy body would fall to earth faster than a lighter one. He thought that velocity is proportional to the moving force and inversely proportional to the resistance, but there was no concept of velocity as magnitude as we understand it today. Motion to Aristotle was a bridge between the potential and the actual.

Aristotle, or his pupil Straton of Lampsakos, wrote the oldest known textbook. It was called *Mechanika* or *Mechanics* and talks about gear wheels, levers applied to weighing balances, and galley oars. It gropes towards an explanation of how a ship can sail into the wind and asks questions about the breaking strength of pieces of wood of various shapes. He used a primitive form of virtual velocities later explained by Hero (1st century AD) as 'the ratio of force to force is inversely the ratio of time to time'.

In 300 BC, Euclid produced his book of geometry called *The Elements*, arguably one of the most important mathematical texts ever written. In it, he brought together many previous ideas and integrated them into a single system of axioms from which theorems could be proved. After Euclid, division lines of considerable accuracy could be drawn on sundials and water clocks. From Euclid to the European Renaissance, geometry was to be the only theoretical language available. In effect, geometry was the mathematical 'spectacles' through which, until Galileo and Newton, people made sense of the world around them. It was the theoretical language of STEM.

One of the first engineers was Archimedes (c. 287–212 BC), although that is not how he is generally remembered and he wrote little of it. Plutarch, in the 1st century AD, wrote:

> Yet Archimedes possessed so high a spirit, so profound a soul, and such treasures of scientific knowledge, that though these inventions had now obtained him the renown of more than human sagacity, he would not yet deign to leave behind him any commentary or writing on such subjects; but repudiating as sordid and ignoble the whole trade of engineering, and every sort of art that lends itself to mere use and profit, he placed his whole affection and ambition in those purer speculations where there can be no reference to the vulgar needs of life....

But this disapproval was from Plutarch (who was no engineer but a well-to-do country gentleman) not Archimedes himself. Indeed, Geoffrey Lloyd says Plutarch may have fabricated it.

Of course, Archimedes is well known as a scientist – chiefly through the story of him jumping out of his bath and shouting 'Eureka'. His famous principle is that a body when immersed in water is subject to an upward force equal to the weight of water displaced – called its buoyancy. We have already mentioned the screw but perhaps his most impressive invention was the compound pulley. Plutarch reported that Archimedes boasted to King Hiero that he could move any weight. So the King and many passengers sat in one of the King's boats and challenged him to pull them along – which he did. Whatever technique Archimedes used to pull King Hiero's boat, he clearly understood the concept of mechanical advantage. Figure 5 shows a modern explanation of how a weight W can be lifted using only one-quarter of W, but Archimedes reasoned about this in an entirely geometric way. Thomas Heath described him as 'the greatest mathematical genius the world has ever seen'. He proved the balanced lever by geometrical symmetry, and his work on the quadrature (finding areas) of curved plane figures gave birth to the calculus of the infinitesimal later perfected by Kepler, Fermat, Leibniz, and Newton.

Geometry enabled craftsmen to formulate 'rules of thumb' for proportioning structures. Most of our knowledge of Roman time is due to Vitruvius; his ten books of architecture were probably written in the 1st century AD. He splits Roman architecture into three parts – the art of building, of making time pieces, and the construction of machinery. He mentioned a number of different types of sundial. The books are like an early engineer's handbook. A rule for the columns of a Forum reads:

> Make the upper columns smaller by one-fourth than the lower, because when it comes to bearing stress, the lower columns should be more substantial than the upper. Do this also because we should imitate the nature of growing things, as in the case of tapering trees.

Vitruvius presented in considerable detail machines such as pulleys, cranes, and methods for pulling large blocks. Of course, military war engines were very important to the Romans. A Roman legion had an *architectus* – master builder; *mensor* – surveyor; *hydraularius* – water engineer; and a *ballistarius* – catapult/artillery-maker.

After the fall of the western Roman Empire in the 5th century, much of ancient learning was protected by the eastern Romans or Byzantines. During this period, Europe was dominated by the Church. The pagan Greek attitude had been that manual work was degrading. The Christian religious view, led by St Benedict and St Augustine in the 6th century AD, was that work was an obligation. That didn't stop the monasteries from using more and more machines to release time for contemplation. The light ploughs used in the drier soils of southern Europe were no use in the heavier soils further north. The wheeled heavy plough had a sharp blade to cut a furrow, a share to slice under the sod, and a mould board to turn it over. Eight oxen were needed but were eventually replaced by the horse. The ox harness pressed on the throat and not the shoulder blades of the horse and so the poor beasts were unable to work efficiently. It wasn't until the 6th century that the breast strap was introduced, and the padded horse collar around the 8th century. Horseshoes were needed in the wet soils of the north. The stirrup was unknown to the Greeks and Romans. Lynn White wrote: 'Few inventions have been so simple as the stirrup, but few have had so catalytic influence on history.' The stirrup gave the horserider lateral support and revolutionized his ability to fight on horseback. It was important in the development of feudalism with a new nobility that challenged the power of the Church. Other improvements included the three-field system of agriculture. One field lay fallow whilst the crops in the other two were used. The next year, the uses were rotated. Yields increased by as much as 50%. The water clocks and water wheels described earlier by Vitruvius were improved, as were windmills, canal locks, and mining. In the UK, by the time of the *Domesday Book* in 1086, 5,624 water mills were recorded south of the Rivers Trent and Severn.

In the meantime, by the 7th century, much of the Mediterranean was taken by the Arab Muslims. The result was an Islamic golden age or renaissance which lasted from around the 8th century to the 13th century and beyond. Centres of learning were established and

Greek texts were copied into Arabic. There was a melting-pot of numerous cultures with important developments in the arts, agriculture, economics, industry, law, science, technology, and philosophy. The library at Cordoba had 600,000 titles by AD 900. Empirical experimentation was not frowned upon. Although religious belief was strong, *mythos* and *logos* were compartmentalized in a way that was to the advantage of practical life. Many of our modern words, such as algebra (Arabic *al-jabr*) and alcohol (*al-kuhl*), derive from this time. The Arabs developed gunpowder and paper (from China), the horse collar, windmill, water wheel, and Arabic numerals. Thinkers such as al-Kindi, al-Biruni, Avicenna, Avernpace, and Averroes introduced Greek ideas into Arabic and developed them. The Muslim thinkers realized Aristotle's physics was inadequate and proposed new ideas which sowed the seeds for work in the 13th century to explain the acceleration of freely falling bodies and the continued velocity of projectiles. For a variety of complex reasons, the Muslim golden age faded and the initiative shifted to northern Europe.

Adelard of Bath had translated Euclid from the Arabic in 1120, though there is evidence that some of the work was known in Europe from the 9th century. No one entering one of the famous Gothic (beginning in 12th-century France) cathedrals, such as the one in Gloucester, could fail to be impressed by the sheer magnitude of the structure. The numerical rules of proportion were formulated as a result of trial and error, taking note of structural success, and perhaps more importantly, of failures. John Fitchen pointed out three-dimensional models were also used during the construction of cathedrals. The architect, the structural engineer, and the contractor were one. Apprentices were trained through the guilds and the more capable became master builders. They were really masters of all phases of the work but, with only a few exceptions, had modest social standing.

Ailnoth (1157–90) was one of the first men to be called an engineer. French was the language of the upper classes in

England, so he was called an *ingeniator*. 'He was a versatile man – a craftsman of technical training…accustomed to making the great war engines which preceded the use of gunpowder.' By the time of the reign of Edward III (1312–77), the men who looked after the firearms were known as *artilators* or *ingeniators*.

The Renaissance marked the beginning of a new era, with an immense change in attitudes. Men such as Brunellesco (1377–1446), Alberti (1404–72), Michelangelo (1475–1564), and Leonardo da Vinci (1452–1519) were typical of the versatile men of this period. They were artist-engineers – people who started studying, learning Latin and mathematics – becoming cultivated. Leonardo da Vinci was one of the first to be appointed as an engineer. He was *Ingenarius Ducalis* (the Duke Master of Ingenious Devices) to the Duke of Milan, and *Ingenarius et Architectus* to Cesare Borgia. Leonardo was the archetype of the Renaissance man. He is one of the greatest painters of all time and his talents covered science, mathematics, anatomy, sculpture, botany, music, and writing. Paolo Galluzzi, Director of the Istituto e Museo Nazionale di Storia della Scienza in Florence, said that between the end of the Roman Empire and before Leonardo the role of the technical worker was generally anonymous. Beautiful buildings were made but the name of the builder wasn't recorded. The intellectual and social distinctions were strong – being trained in the mechanical arts meant that someone who worked with their hands was fit only to work under the direction of someone who was better educated. By the 14th century, people started to use their brains and their brawn.

But still the only theoretical language available was geometry. It is difficult for us in the 21st century to realize that there was no concept of velocity as a magnitude – even by the 14th century. To the Greeks, a magnitude could only come from the proportion of two like quantities such as the ratio of two distances. Velocity and speed involves a ratio of unlike quantities – distance and time. The defect of Aristotle's physics was its failure to deal with acceleration. Thomas Bradwardine and others at the University of Oxford in the 14th century made explicit the ratio of unlike quantities and hence paved the way for velocity and acceleration. They argued that a body moving with uniform velocity travels the same distance as a constantly accelerating body in the same time if that uniform velocity is half of the final velocity of the accelerating body. Nicole Oresme at the University of Paris showed how these ideas could be represented on a graph with time on the horizontal axis and speed on the vertical axis so that distance is the area of a rectangle or triangle. Jean Buridan, also in Paris, suggested the idea of impetus, which he said was the quantity of matter multiplied by its velocity – an anticipation of the modern concept of momentum. Both of these ideas influenced Galileo.

Galileo Galilei (1564–1642) has been called the father of modern science both by Albert Einstein and Stephen Hawking. One of his first pieces of technology was the telescope. He didn't invent the idea, but he did develop it and then proceeded to look at the heavens. The moons of Jupiter weren't fixed but seemed to be orbiting around the planet. Galileo saw what was implicit in the earlier ideas from Oxford and Paris that the distance travelled during a uniform acceleration starting from rest is proportional to the square of the elapsed time. He rolled balls down an inclined plane and timed them using a water clock. He used geometry to conclude that objects move at a given velocity unless acted on by a force – often friction. This was against Aristotle's idea that objects slow down and stop unless a force acts upon them. Galileo stated: 'A body moving on a level surface will continue in the same direction at constant speed unless disturbed.' This was later incorporated by Newton in his first law of motion. Galileo got very close to distinguishing between weight and mass but was unable to make it clear since weight was still seen as an intrinsic downward tendency not depending on an external relationship with another body – an idea that was later to be generalized by Newton in his theory of universal gravitation. Galileo did decide that what persists in motion is the product of weight and velocity which he called *impeto* or *momento* – our modern idea of momentum.

In 1635, Galileo suggested using a pendulum to keep the time, and in 1656 Christian Huygens in Holland built one. These pendulum clocks were much more reliable, so that by the 1700s clocks were beginning to replace sundials. They didn't require sunny skies but often had to be reset from a sundial.

When Galileo was forced to recant, during the Inquisition, his book favouring the Copernican theory that the Sun, not the Earth, was the centre of the universe, he turned his attention to mechanics and published *Two New Sciences*. In it, he considers the tensile strength of a bar, the strength of a cantilever, a beam on two supports, and the strength of hollow beams. Naturally, his solutions are important, but not correct. He assumes, for example, that the stress distribution across the root of the cantilever is uniform, and because he has no concept of elasticity he assumes a constant distribution of stress across the section, right up to the point of collapse. However, he does come to the correct conclusions about the relative importance of the breadth and width of the rectangular cross-section.

Sir Isaac Newton (1643–1727) was the man who really connected time and work. He is arguably one of the most influential men in history. His name is synonymous with classical mechanics. He described universal gravitation and three laws of motion which dominated the scientific view of the physical universe for three centuries. He stated the principles of conservation of momentum. He built the first practical reflecting telescope and developed a theory of colour based on his observation that a prism decomposes white light into the colours of the visible spectrum. He formulated an empirical law of cooling and developed differential and integral calculus at the same time as Leibniz. Newtonian mechanics came to be regarded as the most perfect physical science, and an ideal towards which all other branches of inquiry ought to aspire.

So here at last, we have the relationship between time and work that has served engineering on Planet Earth since Newton and will continue to do so unless we are ever called to build anything that will travel at a speed approaching the velocity of light. From bridges and buildings to aeroplanes and space rockets, Newton's laws are the basis of everything that we have done and much of what we have yet to do.

Three observations are pertinent. Firstly, from pre-history to Newton, science was about trying to make sense of the world around us – it was both *mythos* and *logos*. Between Augustine (5th century) and Thomas Aquinas (13th century), truth became not a reflection of God as much as a relation of things to each other and to man. Their relationship to God was left to theology. Men such as the Venerable Bede showed practical curiosity. By the 12th century, Robert Grosseteste was writing that it was not possible to arrive at absolutely certain knowledge of cause and effect, but it was possible to approach a truer knowledge by making deductions from theories and then eliminating those whose consequences were contradicted by experience. A gradual separation of *mythos* and *logos* had started with the Greeks but was never complete – but nevertheless, the seeds had been sown. After Galileo and Newton, as we will see in the next chapter, science turns into the much more modern idea of 'making sense of the world through systematic rational thinking, observation and experiment in order to understand and make testable predictions'.

Secondly, by this time engineering was emerging as a separate discipline. Many of the great figures from Thales, through Archimedes, to Galileo and Leonardo were driven not only by a need to understand but also to help with practical requirements – military and agricultural. But the separation of technics and science was never as complete as sometimes is supposed. Francis Bacon writing in 1605 said that the mechanical arts had flourished because they were firmly founded on facts and modified in the light of experience. The Middle Ages saw some remarkable technical progress, with new methods for exploiting animal, water, and wind power, inventions such as the mechanical clock and improved magnifying lens. All were in response to clear human needs. The telescope, microscope, thermometer, and accurate clock were later indispensable for the testing of new ideas. The notion that the purpose of science was to gain power over nature was being expressed.

Thirdly, until the 14th century, the theoretical language of mathematics was arithmetic and Euclidean geometry. Only ratios of like quantities were admitted. So, for example, Archimedes, who was arguably the most successful at using mathematics in experimental inquiry, relied on the symmetry of ratios to analyse the balanced lever. Jordanus de Nemore in the 13th century very clearly used virtual displacements but with geometry. The Aristotelian notion of power, velocity, and resistance could not be modelled. As soon as ratios of unlike quantities were used to model velocity and acceleration, a new mathematics of change and motion began. The realization that force was related to acceleration led quickly to Galileo's inertia and Newton's gravitation theory.

In articulating Archimedes' disdain for 'mere trade', Plutarch had reflected the Greek philosophers' (particularly Plato's) attitude that derived from the search for perfection. Two major consequences followed. Combined with monotheism very powerful religions emerged. Science came to be regarded as mere discovery and clearly separated from the individual creativity of art.

But how have these three laws formulated by Newton enabled us to engineer big bridges and send rockets into space? That is the story of modern engineering which we now move on to in the next chapter.

Chapter 3
The age of heat – you can't get something for nothing

Have you ever stretched an elastic band in your hands – then it snaps? Ouch – the recoil can sometimes inflict quite a sharp sting. Before it broke, the band had an internal tension – a pulling-apart force. This force has a capacity to do work, i.e. to recoil and sting you – in other words, it has energy because of the work you did to stretch it.

Before the break this energy is potential – after the break, it has become kinetic. Potential energy is a capacity to do work because of the position of something – in this case, it is because you have moved the ends of the band apart. Another example is a book on a shelf – it has potential energy to fall to the floor. The potential energy in the elastic band has a special name – strain energy. Kinetic energy is due to movement – so if the book is pushed off the shelf or the elastic band snaps, then both lose their potential energy but gain kinetic energy. Kinetic energy depends on mass and speed – the bigger the weight and the thicker the band, the more the kinetic energy.

Whilst kinetic energy is reasonably easy to see, it is perhaps less clear that a bridge, which seems to be a completely static object, has strain energy. All of the various components of a bridge structure have internal forces just like the elastic band though they normally don't stretch quite so much. Also they aren't just tension (pulling) but may be compression (pushing) or shear (sliding) forces as well. These internal forces are caused by the demands on the bridge by the external forces or loads of the traffic as well as its own dead weight and other natural forces such as wind and sometimes even earthquakes. When, rarely, a bridge collapses the internal strain energy, due to all of the internal forces, is released. It is just like the snapping of the elastic band, but on a much bigger scale, of course. All structures, such as buildings, dams, as well as aeroplanes and cars, have internal strain energy just waiting to be released and be turned into kinetic energy should any part of the structure not be strong enough.

But even when a bridge is successfully carrying traffic across a river, it is moving slightly. If you have ever walked on a big suspension bridge such as the Clifton Suspension Bridge in Bristol, UK, or the Brooklyn Bridge in New York, then you will have felt the vibrations and movements as the traffic passes. The structural components of the bridge are responding to the internal forces generated by the traffic just as your elastic band would respond if you moved your hands together and apart to vary the stretch. As a structure responds to the external forces applied to it, so it does work as the forces within it change and the components stretch and compress very slightly.

Everything has structure and hence has some internal strain energy. This is true of a bridge, a building, your home, your car, and even your mobile phone (especially if you accidentally drop it onto the floor). It is even true of public sculptures such as Antony Gormley's contemporary Angel of the North in Gateshead, England. The steel structure is 20 metres tall and has wings 54 metres across and stands on a hill overlooking a major road. Major parts of it had to be designed by a structural engineer to ensure it could stand up safely, especially when the wind blows in such an exposed situation.

So we can see that there are two kinds of work and energy – internal and external. External work is the work done on 'something' – it is a demand. Internal work is work done within something – it is a capacity. When all is well, the internal work equals the external work – but both are constantly changing in a process that is successful only if the 'something' has the capacity to do the internal work required of it by the external work done on it.

At the end of the last chapter, we arrived at the time when Newton formulated laws of gravitation using the new differential calculus. Materials such as iron, steel, and cement started to become available in commercial quantities. The science of engineering quickly flourished in a way that there isn't space here to report in great detail. At first, the new theory had little impact on the practical methods of structural engineering, but by the middle of the 18th century, it was beginning to be useful. For example, in 1742–3 Pope Benedict XIV asked three men, Le Seur, Jacquier, and Boscovich, to find out the cause of serious cracks and damage in St Peter's Cathedral. So they set about assessing the value of the tie force required to stabilize the dome at its base. They postulated a mechanism by which the dome would collapse and calculated the internal and external virtual work assuming it was doing so. They then applied a safety factor of 2 and consequently decided that additional tie rings around the dome were needed, and indeed the work was done. Some people were very unhappy about this new theoretical approach and said:

> if it was possible to design and build St Peter's dome without mathematics, and especially without the *new-fangled* mechanics of our time, it will also be possible to restore it without the aid of *mathematicians*...Heaven forbid that the calculation is correct – for, in that case, not a minute would have passed before the entire structure had collapsed.

The italics are mine – they illustrate an attitude of mind still held today when the gap between new theory (what we think we know) and practice (what we do) seems to get too wide. The gap is an essential part of risk, as we will discover in Chapter 6.

Nowadays, engineers have modern materials such as high-strength steels and composites such as reinforced concrete and plastics. They have computer-analysis tools to help them calculate forces as they design, build, and maintain exciting structures. However, the forces in many existing and complex iconic structures such as the Sydney Harbour Bridge, were calculated entirely by hand. Built in 1932, the summit of the Sydney arch is 134 metres above sea level – a distance that can increase by as much as 180 millimetres on hot days as the steel expands.

Heat is another form of energy. It can be a nuisance in bridges, but it can be made to do useful work by 'heat engines' that turn heat into work. They can take many forms, varying from the internal combustion engine in your car to the giant turbines that generate electricity in power stations. The basic idea is that when you heat up a gas, like air or steam, then its molecules move around quicker and so push harder against anything that gets in the way like a wheel or a propeller (as water strikes a water wheel or air turns the blades of a windmill) or a piston (as in your car engine). They also escape faster through a small hole (as in a rocket – think of a firework). The control of temperature is central to almost all chemical processes that turn raw materials (such as salt, limestone, and oil) into a whole range of chemicals (such as concrete, steel, and plastics) which in turn are converted into consumer products (such as buildings, vehicles, and domestic consumables). Practically all chemical reactions generate heat (i.e. are exothermic) or absorb heat (endothermic). Often, the temperature of the input materials has to be raised to a level required for the chemical reaction and the products cooled for storage – this is called heat transfer and is one of the main disciplines of chemical or process engineering. Heat is also essential in making pharmaceuticals – including synthetic drugs (such as betablockers and antibiotics); food additives and flavourings; agrochemicals such as fertilizers, and petroleum products such as petrol (also called gasoline); diesel and olefins used in textiles, and aromatics used in plastics, dyes, and paints.

Our understanding and use of heat engines was a long time coming – this was because heat puzzled the ancient thinkers. For example, Aristotle argued that quality and quantity were different categories so he concluded that length was a quantity but heat was a quality. He recognized that heat might exist in different intensities but change was not brought about by adding and subtracting parts. He observed that if one hot body was added to another the whole didn't become hotter. In the 13th and 14th centuries, Scotus and Ockham maintained that we could measure heat in numerical degrees but it wasn't until 1714 that Daniel Fahrenheit invented the mercury thermometer with the well-known scale of the same name, followed by Anders Celsius in 1742.

Building on earlier ideas from the 3rd-century Greek inventor Ctesibius, Hero of Alexandria (1st century AD) was one of the first people to use heat to do work – but only to amuse. He distinguished between things that 'supply the most necessary wants of human life' and those that 'produce astonishment and wonder'. For example, he described a way of using fire to open temple doors automatically as well as two dozen other gadgets with secret compartments and interconnected pipes and siphons to produce strange effects. One of his best known is a hollow ball with attached bent tubes with nozzles on the end so that as steam is ejected the ball rotates. This was perhaps the very first demonstration of the potential of the power of steam.

It was another 1,400 years or so after Hero that the potential for heat to do work was again explored, when in 1551 the Ottoman Taqi al-Din described a device for rotating a spit. But it was a very practical need that drove the early development of the first real steam engines – flooding in mines. In 1702, Thomas Savery (1650–1715) built his 'Miner's Friend' to raise water by the

'impellent force of fire'. He may well have seen a 'fire engine'
described by Edward Somerset, the Marquis of Worcester, in 1655.
This first 'engine' had no piston. It worked by heating water in a
boiler by fire in a furnace and piping the steam into a closed
container. The container was connected by a 'suction' pipe down
into the mine water, but a tap or valve closed this pipe off as the
steam entered. Likewise, the container was connected by an
'outlet' pipe to the outside – again with a closed valve. When the
container was full of steam, a valve on the steam inlet from the
boiler was closed so that the steam in the container was entirely
isolated. The steam was then sprayed with cold water and so it
condensed and created a void or vacuum in the container. So
when the valve on the 'suction' pipe into the mine water was
opened the water was sucked up the pipe to fill the void – but it
worked only from a depth of less than 10.3 metres. The suction
pipe valve was then closed and the valve to the outlet opened.
More steam was admitted from the boiler and that drove the mine
water up the outlet pipe to waste. Later versions had two vessels
side by side, with common delivery and suction pipes. In this way,
Savery could have one pumping water out, whilst the other was
filling with water ready to be pumped. Savery compared the work
of his engine with that of horses and so the term 'horsepower' was
born.

In 1707, Denis Papin suggested using a piston, but it wasn't until
1712 that Thomas Newcomen made this work with his atmospheric
steam engine (Figure 7). Newcomen used the same idea as Savery,
a vacuum from condensed steam, but he used it to pull a piston
down a cylinder. The piston was linked to the end of a timber beam
which rotated about a central pin. The other slightly heavier end of
the beam was connected to a pump within the mine water. As the
piston rose and fell, so the beam rotated and the pump rose and
fell – bringing water out of the mine. There were two stages, or
strokes, in each cycle of operation. Starting at the point when the
piston was at the bottom of the cylinder (i.e. at the bottom of its
stroke), there was nothing to stop the heavier end of the beam

pulling the piston up. As it did so, the 'steam valve' into the cylinder was opened and steam flowed in. When the piston reached the top of the stroke the steam valve was shut and a 'water valve' opened so that water flowed in, condensed the steam, and created a vacuum. Then began the second stroke as the atmospheric pressure (hence the name) above the piston pushed it down into the vacuum and pulled the beam with it. As the beam rotated, the other end lifted the water in the pump out of the mine. When the piston had returned to the bottom of its stroke, the condensed water was drained off and the whole cycle started again. At first, the valves were all operated manually, but later this was done mechanically. The engine overcame the height limitations of Savery's engine and hence was able to drain deeper mines.

In 1765, James Watt improved Newcomen's engine by incorporating a separate condenser and then returning the warm condensed water to the boiler. This meant that the main cylinder didn't have to be cooled at each stroke but was kept hot

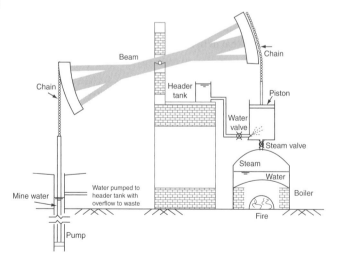

7. **A Newcomen engine**

throughout – a clear increase in efficiency. Watt also sealed the top of the cylinder so that the engine was no longer 'atmospheric' even though there was hot steam above the piston at atmospheric pressure. Later Watt and his partner Matthew Boulton introduced three other innovations. They wanted to use the steam to push the beam up as well as pull it down. This meant that they couldn't use a simple chain to connect the piston to the beam – they had to use a stiff bar. But a bar would not travel vertically up and down as the beam rotated. Watt therefore devised a clever parallel motion mechanism at the end of the beam above the piston so that as the beam rocked the piston moved vertically. In this way, he doubled the power without increasing cylinder size. Second, he devised a centrifugal governor to control engine speed. When the load (the work being done by the engine) reduced, the speed increased, so Watt designed the governor to restrict the steam flow and hence stabilize the speed. Conversely, when the load increased, the speed dropped, but Watt's governor admitted more steam and again stabilized the speed. Watt's third innovation was to design gearing to use a crank (first used by the Romans and later by Al Jazari in the 12th century) to convert reciprocating motion into rotating motion. Around 1800, Richard Trevithick (1771–1833) began to use high-pressure steam to drive a piston rather than condensing the steam to form a vacuum. His engines were called 'puffers' because of the sound made as the steam was exhausted to the atmosphere. As we will see in a moment, this paved the way for smaller efficient engines that could be developed for railway transport.

Despite these clear improvements, the early steam engines were still quite inefficient – a better understanding was needed. Sadi Carnot (1796–1832), a French military engineer who was educated at the École Polytechnique, set about the task – and earned himself the title of the 'father of thermodynamics'. He asked himself whether there was a limit to the number of enhancements that could be made to a steam engine. He looked closely at the way they operated and realized that the cyclic process begins and ends with water. Heat is added to water making steam. Then the steam expands and does mechanical work by pushing a piston. Finally, the steam is condensed back into water. He reasoned that if the inefficiencies arise only from leakages and friction, then he could imagine an engine without them – an ideal engine with a perfectly insulated cylinder and a leak-proof, frictionless piston. As he thought this through, he realized that because the steam had to be condensed to water, then some heat loss was just inevitable – even in this ideal situation.

In this way, Carnot was the first to see the heat engine as a device operating between two heat reservoirs with the steam converting heat energy into mechanical work. At that time, heat was thought of as a flow of a 'caloric' fluid from a hot to a colder body – as water flows from a high level to a low one. Just as the flow, in an enclosed system like a water pipe, is a maximum when the height through which it falls is a maximum, so he said that heat engine efficiency is a maximum when there is the biggest possible temperature difference. Carnot's ideal heat-reversible cycle is one in which heat can be changed into work and work into heat. We now know that it can't be bettered because there are always losses in any practical engine – but we can use it as a comparator.

The Carnot cycle has four reversible stages of expansion and contraction of a gas such as steam – Figure 8 shows how the pressure and volume of the gas change. Two stages are isothermal, i.e. with constant temperature, and two are adiabatic, i.e. with no gain or loss of heat. The full process cycle starts with an isothermal expansion of the gas at temperature T_2 (though in a real engine there would be a temperature drop) shown as a to b in Figure 8 and the piston moves out as the volume increases. This expansion continues into a second stage with the piston continuing to move out but, Carnot reasoned, if this second stage (b to c) is to be reversible then it must be adiabatic, i.e. no heat must be lost as the temperature drops to T_1. Then the piston reaches the end of its movement and the third stage (c to d) starts.

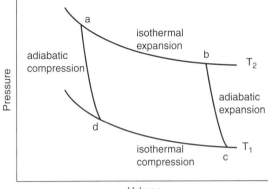

8. **The Carnot cycle**

The piston moves back helped by the inertia of a beam or wheel and compresses the gas. At first, there is no change in temperature, i.e. it is isothermal but the piston continues to compress the gas in the fourth stage (d to a) which again has to be adiabatic if this stage is to be reversible. The total work done in the cycle is the area inside the curve traced out by going from a to b to c to d and back to a. The efficiency of the engine is the proportion of heat supplied from the hot reservoir that is turned into this amount of work. Carnot's theory did not have any significant immediate practical effect but it did provide a datum and it did demonstrate that a heat engine is more efficient the higher the temperature difference within the cycle.

Although Carnot had reasoned all this out successfully using the caloric fluid theory of heat, by the 1840s it had run into great difficulties. Indeed, Carnot himself abandoned it later in his short life – he was only 36 when he died. Sir Benjamin Thompson (Count Rumford) and James Joule both observed effects that the calorific theory could not explain. In 1798, Rumford noticed that the friction generated when boring iron cannons was enough to boil water.

James Joule (see also Chapter 4) generated heat by stirring water. Both Joule in England and Robert Mayer in Germany separately found the amount of mechanical work that is needed to raise the temperature of water by one degree. We now call it the mechanical equivalent of heat, and it is 4.187 joules per gram per degree Celsius.

In 1850, Rudolph Clausius formulated the first law of thermodynamics – that total energy, including heat energy, is always conserved. So when potential energy changes to kinetic energy, then no energy is lost but some may be converted to other forms such as heat. He then went on to state the second law which captures the commonsense notion that you can't get something for nothing – some of the energy becomes irretrievable and no longer available to do work. He said that it is impossible to cause heat to flow from a cold to warmer body unless we supply extra energy. He was the first to show that no engine could be more efficient than the reversible Carnot cycle. In 1865, he coined the term *entropy* to capture the loss of available energy in a heat engine. The total entropy change in a Carnot reversible cycle is zero because there aren't any losses in the ideal cycle. But in an irreversible real process (always less than ideal), then entropy does increase. For example, when heat escapes from a house through a wall, there is an increase in entropy as some of the heat lost becomes irretrievable and unavailable to us to do any work. The idea that in any process some energy becomes unavailable to do work is very important in energy management (Chapter 6).

In Chapter 1, we referred to the way one form of heat engine, the railway steam engine, stimulated social change. The locomotive train became a possibility through Richard Trevithick's engine built in 1804 using steam at a much higher pressure (meaning a few atmospheres, or 30–50 psi). His first train ran from Swansea to Mumbles in Wales in 1807. By 1829, when Stephenson's famous *Rocket* won a competition for the Manchester and Liverpool railway, engines were capable of 30 miles per hour with 30 passengers – Figure 9 shows a full-size replica Rocket built in 2010.

The piston steam engine is an *external* combustion heat engine since the heat source is external to the cylinder. *Internal* combustion engines in modern cars run on hydrocarbons such as petrol (gasoline) and diesel which, unlike steam, have to be made from raw materials by chemical engineering processes. Again, there is a cycle in the four-stroke (i.e. stage) internal combustion engine. It is the fuel and air intake stroke followed by the compression, combustion, and exhaustion strokes (see lower diagram, Figure 11). During the first intake stroke, a mixture of fuel and air passes through a valve into a cylinder deep inside the engine. Then all valves are closed and the mixture is compressed by the moving piston – the second stroke. In the third combustion stroke, a spark from a spark plug ignites the fuel/air mixture which explodes and expands. This forces the piston down and turns a crankshaft. Finally, the exhaust valve opens and the burned gases are released through an exhaust pipe. There are many varieties of internal combustion engine including

9. A modern replica of Stephenson's *Rocket*

two-stroke, four-stroke, six-stroke, diesel, and Wankel engines, as well as gas turbines and jet engines.

Steam piston engines were the dominant source of power well into the 20th century, but have now been replaced by the turbines that generate much of the electricity we use today. A turbine is a rotary engine that extracts energy from a fluid flowing through it. The fluid may be a liquid such as water, or a gas such as steam or air. A water wheel and a windmill are common examples. They have one moving part, a rotor which consists of a shaft with angled blades attached. So when the moving fluid hits the blades, the whole thing rotates. The steam turbine has replaced the piston steam engine because it is much more efficient, with a higher ratio of power to weight. The modern version was invented in 1884 by Sir Charles Parsons. Many power plants use coal, natural gas, oil, or a nuclear reactor to create steam which passes through a huge multi-stage turbine to spin an output shaft that drives an electrical generator. Water turbines are used in hydroelectric schemes to generate power. Water is much denser than steam, so the turbines look different – but the principle is the same. Wind turbines, or windmills, also look different because wind is slow-moving and lightweight.

William Macquorn Rankine (1820–72), a professor of civil engineering and mechanics at the University of Glasgow, Scotland, suggested the cycle that is used now on many electricity-generating plants. In the first stage, the working liquid (generally water) is pumped from a low to a high pressure. The water is heated in the second stage in a boiler at constant pressure by an external heat source to become a 'dry saturated' vapour – which just means that there are no liquid particles in the vapour. The vapour then expands through the turbine and the temperature and pressure of the vapour drop (with some condensation). Then in the fourth stage the vapour is condensed at a constant pressure and temperature to become a 'saturated liquid', i.e. one that contains maximum heat energy without boiling.

Frank Whittle was an officer in the Royal Air Force when he had the idea that he could improve on the performance of the aeroplane piston engine and extend flight distances by using a turbine – by 1930, he had a patent. Similar work was being done in Germany by Hans von Ohain but, despite that, the priorities of the UK government during World War II were elsewhere, hence Whittle had great difficulty in convincing the government that his ideas were worth investment. Consequently, development was slow, but by 1941 a flying version was built. For this, Whittle is often called the 'father of the jet engine' (Figure 10). A gas turbine jet engine has the same four stages as an internal combustion piston engine in your car. It is, however, much more elegant because, rather than happening intermittently, the stages occur continuously and are mounted on a single shaft (Figure 11). The pressure and temperature in a piston engine change quite

10. A jet engine
© Rolls Royce plc 2010

Engineering

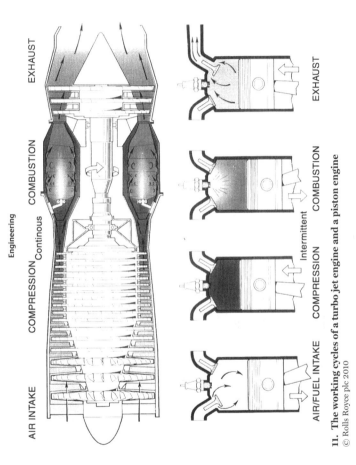

AIR INTAKE COMPRESSION_{Continous} COMBUSTION EXHAUST

AIR/FUEL INTAKE COMPRESSION COMBUSTION EXHAUST

Intermittent

11. **The working cycles of a turbo jet engine and a piston engine**
© Rolls Royce plc 2010

dramatically with time, whilst in a turbine both the pressure and temperature remain constant at steady speeds at given locations in the engine. The gas turbine is a very clever way of manipulating the pressure, volume, velocity, and temperature of gas to create the thrust that propels the aircraft. First, air is taken in (sucked), then it is compressed (squeezed), the fuel burned (bang), and the mix is exhausted (blown) out through a turbine. The engine is a working example of Newton's third law of motion – that for every action, there is an equal and opposite reaction. It's rather like the trick you will probably have done at some time with a toy balloon. You blow it up and release it so that the air rushes out creating the thrust that makes it fly away. In an engine, the air is taken in at the front and expelled at the back. The size of the thrust depends on Newton's second law of motion which states that a force (thrust) is equal to the rate of change of momentum or mass multiplied by velocity. In other words, the thrust depends on the mass of the flow of air through the engine and the difference between the velocities of the air entering at the front (the speed of the aeroplane itself) and leaving at the rear.

An axial compressor in a gas turbine looks rather like a fan but it has a set of specially shaped rotating radial blades, called rotors, mounted on a disc on the central shaft. Alongside each rotor is another set of blades, again specially but differently shaped, called stators, but these are fixed in position and do not rotate. As the air passes through each set of rotors, its velocity increases, and, as it passes through the stators, the gas is diffused turning this kinetic energy into pressure energy – hence the velocity fluctuates but remains essentially the same, whilst the total volume of the gas reduces and the pressure ratchets up by a factor perhaps of the order of 20 to 40 times. As a consequence, the temperature also increases to perhaps 500 degrees Celsius, but the more efficient the compressors, the less the temperature rises. The high-pressure gas then passes along to the next stage – combustion. Here the fuel (propane, natural gas, kerosene, or jet fuel) is injected through a ring and burned. Consequently, the temperature rises dramatically but the pressure remains essentially the same as the volume increases. The hot, high-pressure gas at perhaps around 1,600 degrees Celsius is then accelerated into the turbine by reducing the volume. The turbine has blades like the compressor but shaped differently. The gas is guided by vanes or stators through the rotors expanding as it does and spinning them (essentially a compressor in reverse). The remaining high-pressure gas is then expanded to rush out of the exhaust at high velocity to produce thrust (similar to the balloon). The materials used in the turbine melt at around 1,200 degrees Celsius, so they have to be cooled. This cooling technology applied to a blade made of ice would keep that blade frozen even in the hottest domestic oven. The compressor is mounted on the same shaft as the turbine, and so the turbine spins the compressor at speeds of around 3,000 to 10,000 revolutions per minute at take-off. The forces on the blades as they spin at these high speeds are considerable, and so they have to be specially designed to stop them breaking up.

Modern jet engines in commercial passenger aeroplanes are often turbofan jets. These are gas turbines with a large fan at the front to suck in more air. The thrust of the engine then comes from two sources: the first is the gas turbine itself, and the second is 'bypass air' – so called because it bypasses the turbine portion of the engine and moves straight through to the back of the nacelle (the engine housing) at high speed. The fan may be very big – of the order of 3 metres in diameter – so it can move a lot of bypass air and hence creates much more thrust very efficiently. The speed of the bypass exhaust air is less than that from the turbine, and so the average speed is lower. Since engine noise depends on the speed of the exhaust gases, the turbofan jet engine is quieter.

Probably the most controversial heat engines are those driven by nuclear power. Nuclear electricity-generating power stations work in much the same way as fossil fuel-burning stations, except that a nuclear chain reaction makes the heat. This is then extracted by pumping carbon dioxide or water through the reactor and this, in turn, heats water to make the steam that drives a steam turbine. Nuclear fuel consists of rods of uranium metal, or much more commonly uranium oxides and carbides in metal tubes. Once the process is started, neutrons smash into and split the nuclei of the uranium atoms and heat is released through nuclear fission (splitting the atom). Rods made of boron that absorb neutrons are used as 'control rods' since they can be raised or lowered into the reactor to change the rate of the nuclear reaction.

Fears over nuclear safety were heightened by the Chernobyl disaster in Ukraine in 1986 and by the tsunami damage to the Fukushima Dai-ichi Power Plant in Japan in 2011. Chernobyl was the worst nuclear reactor accident in history. There was a severe release of radioactivity following an explosion after a massive chain reaction had destroyed the reactor. Two people died in the initial steam explosion, but most of the deaths were from radiation. Further explosions and fire sent a plume of highly radioactive gas into the atmosphere which drifted over extensive parts of the western Soviet Union, Eastern and Western Europe. Large areas were badly contaminated, and over 336,000 people had to be evacuated. The incident occurred when one of the reactors was scheduled to be shut down for maintenance – an ideal opportunity to test an emergency safety procedure.

In January 1993, the International Atomic Energy Agency attributed the main cause of the disaster to a poor operational culture and a poorly designed reactor. The reactions to the incident throughout the international community were mixed. For many, nuclear power was simply too risky. It confirmed their view that, although the number of incidents was statistically small, the consequences were too severe to contemplate. Others argued that such slackness in design and operation could not occur in their power stations. Nuclear power can be made safe.

So what of the future? Despite the difficulties, many governments have, perhaps reluctantly, come to the conclusion that nuclear power is a necessary evil as we attempt to cope with the effects of climate change, though many are rethinking their strategy after Fukushima. These and other engineering failures (see Chapter 6) tell us that safety is a complex issue of systems behaviour and, put simply, is a matter of life and death. Nuclear safety brings to a head a central issue within all parts of STEM, albeit in varying degrees – risk. What is the nature of risk? How do we identify the numerous risks inherent in modern engineering and technology? How, in particular, can we remove, remedy, or reduce them to an acceptable level? This is such an important part of all engineering activity, including our approach to climate change, that we will return to it in some detail in Chapter 6.

One possible future technical development of nuclear power that is being actively researched is nuclear fusion – the process that powers the Sun and the stars. It is the reaction in which two atoms of hydrogen combine together, or fuse, to form an atom of helium and some of the mass of the hydrogen is converted into energy. Fusion would be environmentally friendly with no combustion products or greenhouse gases – the actual products (helium and a neutron) are not radioactive. Therefore, with proper design, a fusion power plant would be passively safe and would produce no long-lived radioactive waste. We know the science and we know how to do it – fusion has been done using small particle accelerators that smash deuterium ions into a tritium target at 100 million degrees Celsius and the power is used in nuclear bombs. But the technical engineering challenges are very tough – as yet we are not able to get a sustained, practical, and controlled fusion reaction suitable for the generation of electrical power. Another potentially revolutionary development is an energy catalyser with low-energy nuclear reactions which has been patented by Italian physicist Andrea Rossi but has not yet been adequately tested.

From Hero of Alexandria's steam toys to the NASA space missions, the story of heat and heat engines is one of human *will* driven by practical needs and opportunities to do new things (*activity*) driving curiosity (*knowledge*) to make new forms of motive power and new materials (*objects* and *systems of objects*). The leapfrogging of practical experimentation, production, and science has been tenacious as different people have contributed over many centuries. The mix of skills has also been complex, even within individuals, with some people perhaps more craft-based than others (for example, Thomas Savery and Thomas Newcomen), some tending to be more risk-taking ingenious engineering pioneers and entrepreneurs (for example, James Watt, Charles Parsons, and Frank Whittle), and some more scientific, theoretical, and experimental (for example, Sadi Carnot, Rudolf Clausius, and William Macquorn Rankine). All have contributed insights and moved on our understanding and improved our tools in increments. Progress has not been uniform but stuttering and irregular – but it has been relentless.

In every case, the rigour of engineering and scientific problem solving stemmed from the need to be practical. It required intelligent foresight which includes, but is more than, logical rigour. The whole history of heat engines is not just about finding out new ideas – rather, it is about doing something. The essence of all engineering activity is doing something to fulfill a purpose – a process. It is driven by a strong will to succeed – a will to add value, i.e. to create something of 'worth'. But that process is subject to a whole variety of constraints and that worth will be seen from a number of differing perspectives. Some of those perspectives will conflict – particularly in time of massive social change such as the Industrial Revolution.

The eventual realities – the products, the tools – are not just objects or things. They are complex systems that have a life cycle of their own. So, as we have said previously, they are also sets of processes as they change and have to be maintained – safely. Risk is now a central issue because of the power we now have to change the planet. Safe success requires us to look inwards to deal with detail – 'the devil is in the detail'. The improvements in the very early steam engines (for example, by James Watt) demonstrate how better details, such as valves and control mechanisms, can have a big effect. But engineers also have to look outwards and deal with the 'big picture'. They have to think and reflect as they practise – what Donald Schön has called 'reflective practice'. The many constraints include finance, business, society, and the environment, customer and client needs, as well as operational, maintenance, and regulatory requirements.

But how have heat engines powered the information technology revolution? The answer is by making electricity. In the next chapter, we have to explore the third of the natural processes for doing work – electromagnetism.

Chapter 4
The age of electromagnetism – the power of attraction

Do opposites attract? There is some evidence that dissimilarities in certain genes do affect our choice of partner. Magnets are much more obvious in their preferences – put two unlike poles together (north and south) and they attract. Put two like poles together and they repel.

Unlike heat, electromagnetism (i.e. electricity and magnetism) was a phenomenon only vaguely known by the ancients. They experienced lightning, electric eels, and the effects of static charges just as we do now. Thales of Miletus (640–548 BC) was one of the first to notice that rubbed amber (a fossilized resin) attracts light objects. The ancients knew that lodestone (a naturally occurring form of the mineral magnetite) attracts iron – it served as a primitive compass as long ago as 400 BC. Nowadays, even those of us who know nothing of electromagnetism have probably tried the trick of rubbing a balloon on a rough surface and then making it stick to the wall. Rubbing generates opposite electrical charges on the balloon and the wall and so they attract. You may have noticed that sometimes when you comb your hair or take off a hat, your hairs stand on end – that's because they have developed the same charge and so they repel each other and move apart.

Our modern explanation of all of this is that electric charge is a property of the elementary particles of all known matter. All atoms have a core of various combinations of neutrons (no charge) and protons (positive charge) with one or more surrounding electrons (negative charge). Normally, the number of protons equals the number of electrons, so the atom has no net electric charge. An atom becomes negatively charged if it gains extra electrons and becomes positively charged if it loses electrons. Atoms with net charge are called ions. Particles with electric charge interact with each other through an electromagnetic force. Static electricity has charged particles at rest whereas electric current is moving charged particles. We'll see later the difference between DC (direct current) and AC (alternating current).

The ancients were in awe of some of the strange behaviours that we now know are due to electromagnetic phenomena. John Landels describes how in the 2nd century AD, a ship running from Alexandria to Rome was saved from running aground by a bright star on the masthead. The crew saw this as a divine admonition to turn to port and out to sea. In fact, it was probably a plasma or ionized gas (i.e. one having electrically charged atoms) that glows on sharply pointed objects such as masts of ships at sea during thunderstorms. The phenomenon was named after St Elmo, the patron saint of sailors, who died around AD 303.

According to Hindu tradition, a Benares surgeon of the 6th century BC used magnets in surgery, perhaps for pain relief. In 1269, Peter Peregrinus, a French military engineer, described how to make a compass for navigation. He named the poles of a magnet as north and south, and described the ways they attract and repel. He found that when you cut a magnet in two, you have two new magnets and not two independent poles.

In 1670, Otto von Guericke made the first electrical machine – he generated static electricity (though he didn't realize that was what it was) and man-made sparks from friction between a pad and a rotating glass ball containing sulphur. By the 18th century, electricity was thought to be some kind of fluid. Benjamin Franklin, the American philosopher and statesman, demonstrated in 1752 that lightning is electricity by flying a kite in a thunderstorm. His silk kite, with a wire attached, collected electrical charge from the lightning and carried it down to a Leyden jar. Several people were killed trying to repeat the experiment. Leyden jars, invented in the mid-18th century, were effectively the first batteries or capacitors – devices for holding electric charge. For a time, these jars were the only way to store electric charge.

Charles-Augustin de Coulomb (1736–1806) announced in 1785 the inverse square law relationship between two electrically charged bodies and between two magnetized bodies. Luigi Galvani (1737–98) reported noticing convulsions in the nerves of a frog as he was dissecting it. Alessandro Volta (1745–1827) realized that Galvani had stumbled on the principle of a battery but with biological tissue sandwiched between two metals. So he set out to find alternatives. In 1799, he published his invention of the battery. It had plates of silver and zinc with a piece of cardboard soaked in water or salt water between. This was the first real battery – the first source of a continuous supply of electricity. By 1808, the Royal Institution in London had a monster battery of 2,000 pairs of plates with a total area of 128,000 square inches which is about 80 square metres. It occupied a whole room in the cellar but only produced about 3 kW (kilowatts) – the equivalent of a modern car battery. Inevitably, new improved batteries were developed later, such as the Daniell cell and the Leclanché cell in 1865.

In 1820, Hans Christian Ørsted found that the needle of a compass moved when near to a wire carrying an electric current, and subsequently André Marie Ampère found a way of calculating the magnetic forces due to a current. The first galvanometer to measure current came in 1822, and then in 1827 Georg Simon Ohm announced his now famous law that potential difference V is linearly related to current I through resistance R, i.e. $V = I \times R$.

Michael Faraday (1791–1867) was the pupil of Sir Humphry Davy and his successor as head of the Royal Institution in London. He was a chemist and a great experimental physicist who is dubbed by many 'the father of electrical engineering'. It all started in 1821 when he was asked to write an historical account of electromagnetism for a scientific journal called the *Annals of Philosophy*. He decided to repeat important experiments others had done. Stimulated by his findings, he went on to make two new ingenious devices in which a wire carrying an electric current rotated around a fixed magnet and a free magnet rotated around a fixed conducting wire. In short, Faraday showed that electricity could generate physical work.

He made more remarkable discoveries in 1831. The first was induction, which Joseph Henry (1797–1878) also found independently in the USA at about the same time. Faraday showed that a steady current in a conducting coil had no effect on another similar coil, but a changing current created a changing magnetic field or zone of influence (we will see how the idea of a field developed in a moment) and induced a current in the other loop. Induction happens when an electrical charge produces a magnetic or electrical effect in another body *without any direct contact*. Within a month, he built the first electric generator – the 'Faraday disc' of copper that turned between the poles of a powerful magnet. He had demonstrated that mechanical work could be converted into electricity.

No sooner had he done this, than the French instrument-maker Hippolyte Pixii made a 'magneto-electric machine' in 1832 that produced a somewhat discontinuous alternating current (AC), i.e. one that varies in direction. He put a coil of wire over each pole of a horseshoe permanent magnet and an iron core within each coil.

He then rotated the magnet under the coils using a hand crank and thus generated an electrical current in the coils. However, as the magnet rotated so the direction of its movement with respect to the coil changed every 180 degrees. As a consequence, the direction of the current also changed, and so he got an alternating output. At that time, direct current (DC) was much preferred, so he then used a see-saw arrangement, devised by Ampère, as a changeover switch to make the current always flow in one direction, i.e. to make direct current. He had developed a very early version of a commutator – a device for reversing the direction of a current.

Soon afterwards, in London, Joseph Saxton fixed the heavier magnet and instead rotated the lighter coils containing the iron bars. Figure 12 shows just one rotating coil and the wavy form of

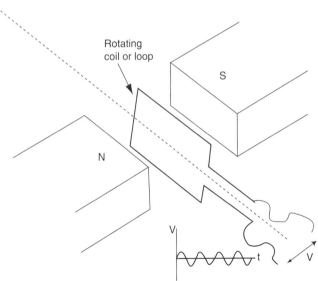

12. A rotating coil

the voltage V produced through time t. He and E. M. Clarke in 1834 developed probably the first commercial hand-cranked generators. By 1862, two large machines developed by F. H. Holmes were driven not by hand but by steam engines and they were used to power arc lamps at Dungeness lighthouse in the UK.

Large-scale power was still not feasible, however, until the permanent magnets were replaced by electromagnets. Electromagnets are not permanently magnetic – they are iron bars that only become magnetic when an electrical current is passed through wire coils wound around them. The wire coils are called field windings. The clever part was that the DC generator (dynamo) was self-excited, i.e. some of the current being generated was also used to create the electromagnets. It worked because the iron bars retained some residual magnetism – just enough to enable the dynamo to produce some output voltage, causing current to flow in the field windings and fully magnetize the bars.

Werner von Siemens produced the first big improvement in the efficiency of dynamos when in 1856 he developed an iron cylinder armature (the rotating part) with slots containing the field windings (see Figure 13). The AC was converted to DC using a split ring commutator to give the output voltage V as it varies through time t as shown in the figure. Soon after, in Italy in 1860, Antonio Pancinotti produced a machine with a solid ring armature that could be used as a generator (i.e. converting mechanical to electrical energy) or a motor (i.e. converting electrical to mechanical energy). About 10 years later, a Belgian working in Paris, Zénobe Théophile Gramme, used Pancinotti's idea of a ring armature, but instead he formed it from iron wire. He obtained, for the first time, a more or less continuous DC by tapping the wire at very short intervals around the ring and connecting them to a multi-segmented commutator. In doing so, he effectively smoothed out the humps in the variation of voltage V through time t in Figure 13, and the greater the number of segments the smoother the DC. His machines were improved and made in several countries and used mainly for lighting. Siemens in Germany and England, R. E. B. Crompton in England, Thomas Edison in the USA, and many others contributed applications in factories, agriculture, and locomotives – the first electrical train was opened in the 'deep' London Underground in 1890. A new era in the generation and use of electricity was ushered in – electrical engineering moved from infancy to adolescence.

Up to this time, DC power was preferred, but unfortunately it could only be supplied over relatively short distances. AC generators were being made but could not be widely used because it was difficult to synchronize the wave form of their outputs. They were therefore only operated singly. The real breakthrough for AC came when in 1888 Nikola Tesla, an Austrian-Hungarian who had worked for Edison, announced that he could make a rotating magnetic field from AC currents which are out of step. His

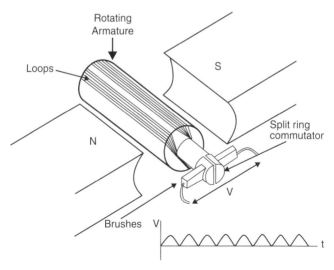

13. A DC generator

induction alternator consisted of an electromagnet (the rotor – shown in Figure 14 just as a bar magnet for simplicity) rotated, by an external source, within several separate pairs of coils of copper wire – in Figure 14 just two pairs of single wire coils are shown set 90 degrees apart. The rotor magnet creates a magnetic field which, as it rotates, cuts across the wires in the stator and creates an alternating electrical current (AC) in each of the pairs of coils. Consequently we get two AC currents with the same frequencies as shown but with peaks and troughs set 90 degrees apart. A three phase alternator (not shown) has, in its simplest form, an iron stator with three sets of pairs of coils or armature windings, each physically offset by 120 degrees (i.e. a third of a circle) and a rotor magnetized by field windings to create three AC currents set 120 degrees apart, as shown bottom right in Figure 14.

A motor is a generator in reverse, i.e. it converts electrical energy into mechanical energy. In a modern AC induction motor, there is no physical connection between the rotor and the stator (shown separately in Figure 15). Instead when a three-phase voltage is applied to three sets of coils in the stator (stator windings) a rotating magnetic field is created which induces current in rotor bars. These currents, in turn, produce a magnetic field, which reacts with the rotating field, as the like poles repel and the unlike ones attract, to produce torque – and hence rotation. In effect, the rotating magnetic field drags the rotor around. AC induction motors are simple, robust, elegant, unglamorous workhorses driving untold numbers of pumps, fans, compressors, hoists, and other modern machinery.

The number of AC cycles per second are called Hertz – abbreviated to Hz. George Westinghouse in the USA began to promote three-phase 60-Hz AC distribution systems, but Thomas Edison (1847–1931) clung to his belief in small-scale DC. He assumed industrial companies would build their own generating plants with his parts. He didn't see the next step into large power plants and the creation of a national grid to share

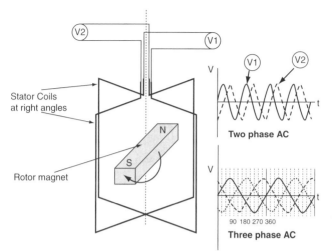

Stator Coils at right angles

N

Rotor magnet

S

V1 V2

V

t

Two phase AC

V

t

90 180 270 360

Three phase AC

14. Tesla's two-phase AC

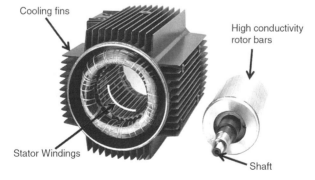

Cooling fins

High conductivity rotor bars

Stator Windings

Shaft

15. A modern AC induction motor

power. Eventually, however, Edison had to switch to AC. The reasons were that in a DC system the voltage must be generated at the level required by the load. For example, a 240-volt lamp must have a 240-volt DC supply. If the same generator is used to supply a 120-volt lamp, then a resistor or another 120-volt lamp has to be in series with the 120-volt lamp to drop the extra 120 volts – this is wasteful. Power losses in a conductor are a product of the square of the current and the resistance of the conductor. So the same amount of power can be transmitted with a lower current and a higher voltage and lower losses. Of course, there are disadvantages in using high voltages. The main one is the need for more insulation and the difficulty of handling them safely. But the development of the transformer was pivotal for AC. Power could be generated at a convenient voltage and then stepped up to a high transmission voltage. At the receiving end, near the electrical demands or loads, the voltage could be stepped down to that of the equipment. These voltages vary between countries and size of load. Typical figures are 110 volts in the USA and 240 volts in the UK.

At first, these new alternating currents raised considerable fears. Indeed, Edison tried to magnify those fears through a long-fought PR campaign in which animals were executed in public using AC. Death-row prisoners were executed for the first time using AC in 1890. But Edison moved on, and the man who eventually delivered AC for him was an Englishman called Samuel Insull. He had the job of running Chicago Edison Company, a small independent power supplier with 5,000 customers. Insull saw that large central stations would be able to deliver power for a lower price than individual manufacturers. He expanded the capacity of his company from being the smallest in 1892 (1 of 12 in Chicago) to one with 500,000 customers in 1920.

In 1882, Sebastian Ziani de Ferranti set up a shop in London designing electrical devices. He bet on the success of alternating current power distribution early on, and was one of the few experts in this system in the UK. In 1887, the London Electric Supply Corporation (LESCo) hired Ferranti to design their power station at Deptford. In 1891, this was the first modern power station that supplied high-voltage AC power that was then stepped down to a lower voltage for consumers. This basic system remains in use today around the world.

Arc lamps were first invented by Davy in the early 1800s. By 1860, Joseph Wilson Swan was on the way to producing an incandescent light bulb. He used carbonized paper filaments in an evacuated glass bulb. Unfortunately, he could not get a good vacuum and an adequate supply of electricity so the bulb had a very short lifetime. Nevertheless, by 1881, he had taken out a British patent, installed electric light in his own home and established his own company. Meanwhile, in the USA, Edison in 1878 took his first patent for an electric bulb based on Swan's UK patent. He set out to replace gaslights with electric ones. He reasoned that electricity was easier to control than gas. Gas sucked oxygen from rooms and gave off toxic fumes, blackened walls and soiled curtains, heated the air, and sometimes caused explosions. Electricity was cleaner and safer. But he had to find ways of producing electricity efficiently in large quantities and then transmit it safely to homes and offices. He needed ways to measure how much each customer used and to turn the current into controllable, reliable light suitable for a home – all at a price that could match that of gaslight and turn a profit.

Part of his solution was the Edison light bulb. It had a sealed thin copper filament inside a small glass vacuum. His company designed a whole system with a new dynamo and a parallel circuit to allow many bulbs to operate independently with separate controls on a single wire. They developed and marketed all the bits and pieces for a complete distribution system: underground cables, electric meters, wiring, fuses, switches, and sockets. By 1882, Edison had installed three 125-horsepower 'Jumbo' generators at the Pearl Street Station in New York, which fed power to 5,000 lamps in 225 houses. By 1895, electricity was widely available in commercial sections of large cities.

In the 1890s, transmission by AC began to cover large geographical areas because generators no longer needed to be close to where the power was needed. A new type of engineer emerged – the electrical engineer. The first training in the UK was the School of Telegraphy and Electrical Engineering in London.

World War I created huge demands for electricity but the industry had become fragmented. By 1918, in London alone, there were 70 authorities, 50 different types of systems, 10 different frequencies, and 24 different voltages. In the UK, an Act of Parliament in 1926 created the Central Electricity Board which set up a national AC grid, running at 132 kV and 50 Hz. By 1933, there were a series of interconnected regional grids which were operating as a national system by 1938. In 1949, the grid was upgraded with some 275 kV links, and again in 1965 with some 400 kV links. The UK grid was nationalized in 1947 and denationalized in 1989. In the USA, the Continental power transmission grid consists of about 300,000 kilometres of lines operated by approximately 500 companies.

National grids in all countries transmit electrical power derived from mechanical generators that are mainly heat engines – usually steam turning a turbine as described in the previous chapter. Steam is generated by burning coal, oil, natural gas, or other combustibles such as biomass (plants) or from a nuclear fission reaction. Natural gas and petroleum are also burned in gas turbine generators where the hot gases produced from combustion are used to turn the turbine. However, any source of mechanical power that can rotate the armature inside a generator can be used – for example, windmills or hydroelectric schemes where moving water drives water turbines or water wheels. Tidal power depends on water flow to turn low-pressure head turbines. Combined heat and power (CHP) is the simultaneous generation of heat and power. In its simplest form, a gas turbine, an engine, or a steam turbine drives an alternator and the heat produced is recovered and used to raise steam for industrial processes or to provide hot water. CHP systems make use of the heat produced during electricity generation with overall efficiencies in excess of 70%. This is in contrast with the usual efficiencies of conventional coal-fired and gas-fired power stations, which discard this heat, of typically around 38% and 48% respectively.

Some renewable energy systems rely on a local distribution network of smaller generators with power bought and sold to the national grid to supplement centralized power stations. So fairly small-scale power generation – typically between 3 and 10 kW – can provide useful supplementary power based on, for example, photovoltaic solar panels, small-scale wind turbines, and Stirling engines. Photovoltaic cells are made from two or more layers of semiconducting material such as silicon. The light generates electrical charges that can be conducted away as direct current (DC). The output from a single cell is small, so many cells must be connected together to form modules or panels. The equipment has no moving parts and therefore is silent, with minimal maintenance and no emissions of greenhouse or any other gases.

New and innovative ideas for creating renewable energy depend on the way we understand how energy processes work. For example, future medical sensors may be able to harvest power from the warmth of the human body. At the time when Faraday started his research, electricity and magnetism were conceived as fluids. But his genius and ingenuity, combined with extensive experimentation, led him to an intuitive notion of a field. A field is a region of space under the influence of a physical agency such as electricity, magnetism, or gravitation. A charge or a mass in a field has force acting on it, so a charge in an electromagnetic field experiences both electric and magnetic forces. He conceived the idea after noting the pattern assumed by iron filings near a magnet. His intuition was that all electromagnetic forces were distributed in well-defined geometrical patterns.

James Clerk Maxwell (1831–79) developed these intuitions into field theory. Maxwell was certainly one of the great theoretical physicists of all time. His equations generalized much of what Faraday had discovered about electricity and magnetism. Both Faraday and Maxwell conceived of lines of force as perturbations of the space surrounding charges, magnets, and currents, i.e. as a state of that space. Through Maxwell's work, we began to understand light as an electromagnetic wave in a spectrum ranging from the longest, which are radio waves with frequencies of around 10^5 Hz or lower and wavelengths expressed in kilometres and greater, down to the very short gamma rays with frequencies of 10^{20} Hz and wavelengths expressed in nanometres (10^{-9} metres). In between, there are microwaves, infrared waves, the visible region or light, ultraviolet light, and X-rays. Maxwell's equations were a turning point in the history of science on which many others, including Albert Einstein, were later to build theories of relativity and quantum mechanics. It was, as Thomas Kuhn was later to point out, a scientific revolution.

However, at the time, few saw it this way. Even the famous Sir William Thomson (1824–1907), who later became Lord Kelvin and who was Maxwell's mentor, dismissed Maxwell's theory as 'curious and ingenious, but not wholly tenable'. In 1904, just before he died, Kelvin maintained that 'the so-called electromagnetic theory of light has not helped much hitherto'. Nevertheless, Thomson's fame had been well earned – he was a great physicist and engineer. In 1851 to 1854, building on work by others, Thompson restated the two laws of thermodynamics (see the previous chapter). The first – the law of equivalence or conservation of energy – he attributed to Joule (1843), and the second – the law of transformation – he attributed to Carnot and Clausius (1850). He suggested the absolute scale of temperature (degrees Kelvin) and introduced the idea of available energy, which is central to the concept of entropy (see Chapter 3). From 1881, he took greater interest in the generating problems of electrical engineers – particularly the accumulator or storage battery invented by Camille Alphonse Faure (1840–98). He collaborated with Ferranti in the design of a special winding for an AC dynamo and was a consultant to Ferranti and Crompton. He led an initiative by the British Association for the Advancement of Science to establish a common set of units on which the development of engineering science could be based.

However, it was Maxwell's approach that presaged modern scientific thinking. By this view, the details of the basic structure of our world are beyond common sense, and we must rely on mathematics as an abstract source of understanding. Of course, the success of any way of understanding the world around us depends on how it responds to tests. If a theory predicts something new, such as a different kind of behaviour, then we must search for it and if we find it then a test is passed. Nevertheless, no matter how many tests are passed, there may still be a test in the future that our theory will eventually fail – we can never be sure. Maxwell's theory immediately passed some major tests. The idea of electromagnetic radiation in Maxwell's equations led to the discovery of radiation at other wavelengths of which radio, radar, and television are the prime examples.

Indeed, radiation was about to revolutionize the way we humans send messages to each other. Naturally, people have always wanted to send messages. Over 3,000 years ago, carrier pigeons were used. People have used smoke, fire, beacons, and semaphore. It was therefore completely natural to try to use electricity. The first attempts at telegraph in the early 1800s used 35 wires, one for each letter of the alphabet and numerals, and succeeded in sending messages a few kilometres. Each wire was dipped in a separate tube of acid at the receiving end. When current from a battery was switched in to a particular wire at the sending end then a stream of bubbles was released at the receiving end – revealing which letter had been sent. The first regular telegraph was built in 1833 when the signal was detected using a galvanometer to measure current. Samuel Morse took out a patent for his famous system based on dots and dashes in the USA in 1837. The first successful commercial application was in 1839 when signals were sent from London Paddington railway station to West Drayton 13 miles away. In 1865, a cable being laid across the Atlantic was lost. William Thomson (Lord Kelvin) was knighted for his work on helping to reconnect it. He was the ruling spirit behind successful work to recover 1,000 miles of cable in 1866 when a new cable was laid and the lost one recovered. He patented a mirror galvanometer in the UK in 1858 that was sensitive enough to detect variations of the current in a long cable. It could detect a defect in the core of a cable. It was, at that time, the only practicable method of receiving signals over long-distance cables.

Thomas Edison in the USA devised a way of sending two and then four messages down a single cable in 1874. By 1902, a cable was laid across the Pacific Ocean and the world was encircled. But at the same time, people were beginning to explore wireless telegraphy. Guglielmo Marconi (1874–1937) transmitted one of the first wireless signals over 6 kilometres in 1896 – it was still Morse code, but it was also the beginning of radio. He failed to interest the Italian government and so came to England, where he eventually succeeded in sending the first radiotelegraphy (telegraph without wires) transmission across the Atlantic in 1901. In 1909, he and Karl Ferdinand Braun shared the Nobel Prize for Physics for 'contributions to wireless telegraphy'.

In 1888, Hertz had produced and detected electromagnetic radiation using a spark-gap transmitter and a spark-gap detector located a few metres away. A spark-gap transmitter is a spark gap (i.e. a space between two electrodes) connected across a circuit with a capacitor and inductor to form a simple oscillator. Today, we call this an LC oscillator – it is the electrical equivalent of a mass vibrating on the end of a spring. L stands for an induction coil of wire which stores energy as an electro*magnetic* field. C stands for a capacitor – Hertz used a Leyden jar – which also stores energy but as an electro*static* field which creates a potential difference. When the capacitor has enough charge, its potential difference causes a spark to form across a spark gap, forming electromagnetic waves (which Hertz observed were being picked up by the detector). A current is also created in the wire induction coil with its own electromagnetic field. As a consequence, the potential difference of the capacitor reduces and the capacitor starts to discharge and lose energy but, at the same time, the field of the induction coil strengthens and it starts to collect energy. Eventually, the capacitor is completely discharged with no current through the coil and so its field strength begins to reduce. This creates a current that flows back through the capacitor, recharging it – but with opposite polarity. Once the capacitor is totally recharged, the whole process starts over again. In this way, the circuit oscillates with a frequency that depends on the characteristics of the capacitor and the induction coil. Some energy is lost in each cycle through resistance in the circuit and heat at the spark gap, and so the oscillations decay or are damped. In 1893, Nikola Tesla used a coil in the detector or receiver tuned to the specific frequency used in the transmitting coil. He showed that the output of the receiver could be greatly magnified if it could be made to resonate (see Chapter 2). In 1898, he patented a radio-controlled robot-boat which he demonstrated that year at the Electrical Exhibition in New York.

At the same time, and after reading the work by Hertz and the Russian scientist Alexander Stepanovich Popov, Marconi set about some experiments using a long pole to pick up radiating electromagnetic waves. The long pole became known as an antenna (Italian for pole) or, more commonly in the UK, an aerial. The antenna converts electromagnetic waves into electrical currents when receiving a signal and vice versa when sending or broadcasting. A common version is a vertical rod one-quarter of a wavelength long – perhaps 2.5 metres for a 30-MHz signal – in which the electrons resonate with maximum amplitude as the difference in voltage (between a node and an antinode over one-quarter of the wavelength) is at a maximum.

As Marconi was developing the radio, Alexander Graham Bell (1847–1922) wanted to improve the telegraph, and in doing so invented the telephone. The telegraph was a very limiting way of communicating. Bell wanted to find a better way of transmitting multiple messages over the same wire at the same time. His knowledge of music helped him to conceive the idea of an harmonic telegraph whereby several notes of different pitch could be sent at the same time. By 1874, Bell had advanced his ideas sufficiently to persuade his future father-in-law to fund him. The following year, he had demonstrated that different tones would vary the strength of an electric current in a wire. He just needed to build a transmitter capable of producing varying electrical currents in response to sound waves and a receiver that would reproduce these variations in audible frequencies at the other end of the wire. In 1876, Bell spoke down a wire to his assistant, Thomas A. Watson, in the next room, 'Mr Watson – come here – I want to see you.'

A telephone is one of the simplest ways of processing information. It has a switch to connect and disconnect the phone from the network, a microphone, and a speaker. In general terms, a microphone is an instrument that responds to varying sound pressure waves in the air and converts them into varying electrical signals. The speaker does the reverse. So when you speak into a telephone the sound waves of your voice are picked up, in the simplest microphone, by a diaphragm combined with some carbon granules or dust. As the diaphragm vibrates, it compresses the dust which changes resistance and hence varies the electrical current that flows through the carbon. In a dynamic microphone, when the sound waves hit the diaphragm, either a magnet or a coil is moved and a small current created. A speaker takes the electrical signal and translates it back into physical vibrations to create sound waves. When everything is working as it should, the speaker produces nearly the same vibrations that the microphone originally received.

In 1877, Edison attempted to put telegraph messages onto paper tape by putting indentations of various sizes, corresponding to the message, in the paper. That led him to wonder if he could capture the sound waves of a voice from a telephone message in a similar way. He experimented with a diaphragm and a needle pointer to make indentations, corresponding to the vibrations of the sound of the voice, in moving stiff paper. Then, he reasoned, he could play the indentations back by running a needle over them. Later, he changed the paper to a metal cylinder with tin-foil wrapped around it. He designed two diaphragm-and-needle units, one for recording and one for playback. John Kreusi built the machine to Edison's design. It is said he immediately tested the machine by speaking the nursery rhyme into the mouthpiece, 'Mary had a little lamb'. And to their amazement, the machine played his words back. Recorded sound was invented.

Soon Edison and his competitors were using wax cylinders and then discs. By around 1910, discs of shellac were being played at 78 revolutions per minute (rpm) and these were followed by LPs, or long-playing vinyl records, at 45 and 33 1/3 rpm in 1948. In all of these, a groove with many bumps and dips is cut or pressed into the disc. A needle running along the grooves vibrates as it strikes the bumps and dips. A magnetic pick-up converts the vibrations into an electrical signal which is carried by wire to an amplifier and speakers. Vinyl records continued to be a major form of recorded media until the mid-1980s, when the audio tape took over, and eventually gave way to CDs and MP3 players and digital files downloadable from the internet.

Reginald Fessenden (1866–1932) was the first to attempt to transmit the human voice by wireless, in 1900. Like Hertz, he used a spark transmitter but one making around 10,000 sparks/ second. He modulated the transmitted signal using a carbon microphone – in other words, he combined the signal from the spark transmitter with the signal from the microphone, but the sound it made was hardly intelligible. The problem was that spark-gap transmitters generate decaying signals over a broad range of frequencies. What was needed was a well-defined continuous oscillating wave. It eventually came via the vacuum tube, first invented by the Englishman J. A. Fleming in 1904, and perhaps better known as a valve.

A vacuum tube is rather like an incandescent light bulb – but one that can switch, amplify, or otherwise modify an electrical signal. As a switch, it operates like a water tap or valve that can turn flow on or off – hence the name 'valve'. The simplest vacuum tube is a diode with a filament cathode (negative) and a plate anode (positive). When the anode is positive with respect to the cathode, then the electrons move easily from one to the other – this is called forward bias. When the voltage is the other way round, it is hard for the electrons to escape from the cathode and the flow is reduced almost to zero – this is called reverse bias. So effectively, diodes only allow current to flow in one direction.

Another kind of vacuum tube is called a triode, first developed in the USA by Lee De Forest in 1907 after disputes with Fleming and Marconi about patents. It is used to amplify a signal by taking energy from a separate power supply. As the name implies, the triode has three electrodes – the cathode and anode as for the diode but with a grid interposed between. The grid is a fine mesh usually near to the cathode. Two voltages are applied – one a power supply between the anode and cathode, and the other an input signal between the grid and cathode. As a consequence, an electron is influenced by two separate electric fields. But because the grid is close to the cathode, the electrons are more sensitive to changes in the input grid voltage than the power supply anode voltage. Thus, a small change in the input grid voltage can produce large variations in the current from cathode to anode and therefore (using the valve in a suitable circuit) the output voltage at the anode. Depending on the size and closeness of the grid to the cathode, the voltage change can be amplified 10 to 100 times.

The vacuum tube was used as an amplifier as part of an oscillator to provide the well-defined oscillations needed to form a carrier wave which, as we will see in a moment, was essential for the development of radio, television, radar, telephones, computers, and industrial control processes. It worked by feeding back the output from the amplifier in phase with the input in such a way that maintains the oscillation without any damping and at a controllable frequency which depends on the characteristics of the amplifier and the feedback circuit.

Radio signals are transmitted via analogue (continuous) carrier waves which are modulated (as we said earlier, this just means combined, changed, or modified) by an input signal or wave. The input signal contains the information we wish to transmit, such as that from someone speaking into a microphone. The two main ways this was done are AM, or amplitude modulation, and FM, or frequency modulation. We will see how digital systems work in the next chapter. The modulated signals – AM or FM – are detected by a radio receiver tuned to that particular frequency by turning a knob that varies the capacitance of the capacitor until the receiver circuit resonates. The signal is then demodulated or decoded to reproduce the original sound, which is then amplified and sent to a speaker.

In AM, the carrier wave is modulated in proportion to the amplitude or strength of the input signal so the carrier wave varies as shown in Figure 16. Figure 17 shows an AM signal for the spoken word 'six'. The earliest AM technique was the Morse code, where the amplitude is a maximum when a key is pressed and zero when released. As we saw earlier, Fessenden had tried to modulate the signal from a spark-gap transmitter with output from a microphone, but the output was barely audible. From 1922, early radio broadcasts by the BBC were AM. Although AM is simple and robust, there is quite a lot of interference from household appliances and lighting, with night-time interference between stations. It also requires high battery power.

Edwin Armstrong (1890–1954) developed FM radio in the 1930s. In FM, the frequency rather than the amplitude of the carrier wave is varied, as shown in Figure 16. The BBC started FM transmissions in 1955. FM waves deliver good voice quality, are less susceptible to interference, and are able to carry more information. However, the higher frequencies need a line of sight and so are interrupted by large obstructions such as high hills.

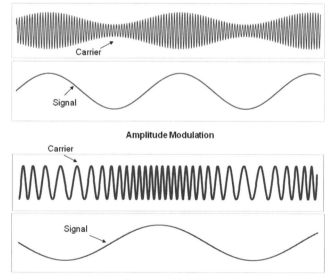

Amplitude Modulation

Frequency Modulation

16. **AM/FM modulation**

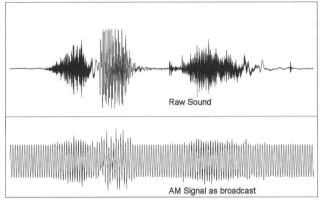

Raw Sound

AM Signal as broadcast

17. An AM signal for the spoken word 'six'

The lower frequencies used for AM can travel further distances as they are reflected back from the ionosphere, whereas FM passes straight through. FM allows more bandwidth – the smallest range of frequencies within which a particular signal can be transmitted without distortion.

Electromagnetic work was being used to send information, but it wasn't long before it was beginning to be used to process information, and it is that to which we turn next.

Chapter 5
The age of information – getting smaller

When I was a student in the 1960s, my university was very proud of its computer – a Ferranti Mercury that occupied a very large room with a dozen or so wardrobe-sized cabinets full of valves, capacitors, resistors with miles of wire, and thousands of soldered joints. Now I have orders of magnitude more processing power in the mobile cell phone in my coat pocket. The miniaturization of electronic processing power over the last half of the 20th century has been remarkable and shows no sign of abating in the 21st century as we pursue nanotechnology and quantum computing.

The seeds of what many are calling the new 'industrial revolution' – the ongoing information revolution – were sown when we began to transmit information using our newfound discoveries about electromagnetism in the early 20th century. In this chapter, we will start by looking at the icon of this new age, the mobile cell phone, and see how it relies on the revolutionary invention of the transistor. We will then briefly examine the subsequent miniaturization of components and how they have been incorporated into digital equipment such as computers.

A mobile cell phone is actually a two-way radio rather than a traditional phone and is for many also a personal digital assistant that includes a camera, an internet connection, email, music, and video, as well as other services such as maps and GPS (global positioning systems). The way the phone works is ingenious. A region, such as a city, is divided into small areas, called cells, each of about 25 square kilometres. Each one has a base station with a tower and radio equipment that connects to the phones. Your operating company may have regulatory approval to use something like 800 radio frequencies for a big city – not enough for all of our phones to have one each. The transmitter in your phone is low-powered, which saves on batteries, but also means it cannot send signals much further than one cell. This has the advantage that the phones can then use the same frequencies in different cells as long as they aren't adjacent.

Each company has a mobile telephone switching office (MTSO) where all the calls in the region are received from the base stations and are passed on to the normal phone system. Each mobile cell phone has special codes which identify the phone, the phone's owner, and the service provider. So when you switch on your phone, it connects, using a control channel with a special frequency, to its home base station and hence to its MTSO. The MTSO keeps track of the location of your phone using a computer database. When the MTSO gets a call intended for you, it looks to see which cell you are in. It then picks a frequency pair for your phone to operate with the base station in your cell, and you are connected with the MTSO via that base station. If, as you are taking your call, you move towards the edge of your cell, then your base station notes that the strength of your signal is reducing and the base station in the cell you are moving towards notes that it is increasing. So the two base stations coordinate through the MTSO, and at a suitable moment your phone gets a signal on the special frequency control channel telling it to change frequencies into the new cell. So, as you travel, your signal is passed from cell to cell, and you don't notice. In this way, millions of us can use our mobile phones simultaneously. Soon, femtocells will bring new and even faster coverage, with new and better services.

The mobile cell phone has only become possible through the miniaturization of electronics. It all started when semiconductors and the transistor were invented – work for which William Shockley, John Bardeen, and Walter Brattain were awarded the Nobel Prize for Physics in 1956. Transistors have effectively replaced vacuum tubes. As amplifiers, they enable a small voltage to control the flow of a much bigger current – just as a valve controls the flow of water in a pipe. The components are very small, very pure, and solid, with no moving parts, and consequently are much more robust. They are used individually, but they are most often found in the integrated circuits that make nearly all modern electronic equipment. We shall look at their role in digital devices such as logic gates and flip-flops that drive microprocessors and computers in a moment.

There are many types of transistor depending on the material used, its structure, power rating, operating frequency, amplification level, and intended application. The two most common are the bipolar transistor and the field-effect transistor. Transistors are made of materials such as silicon and germanium which are not good conductors like copper but can be 'doped' in a way that changes their conductivity – hence the name 'semiconductor'. One dopant in silicon is phosphorus – an *n-type* – because it has more negatively charged electrons than the silicon needs to bond to it. It therefore effectively donates free electrons (Chapter 4) to the semiconductor. Boron is one example of a *p-type* dopant because it has fewer electrons than the silicon requires to bond to it. It therefore steals them from the silicon, and this effectively creates 'holes' of positive charges in the semiconductor which attract any negatively charged electrons that happen to be around.

Transistors are so very important that it is worth looking at how one type, the bipolar transistor, works in some detail. It was the first to be mass produced and is bipolar because both negative electrons and positive 'holes' move. It also has two circuits – the base and the collector, as shown in Figure 18. Note that the arrows showing the direction of flow of the current are, by convention, positive to negative and hence opposite to the direction of flow of the electrons. The transistor consists of semiconductors used in wafer-thin layers, rather like sandwiches, with *n-type* as the 'bread' and *p-type* as the 'filling' (*npn*) – as in Figure 18 – or the other way round (*pnp*). The transistor acts like a triode (Chapter 4) of two diodes that have two cathodes but share a common anode. The equivalent of the two cathodes is the 'bread' in the transistor sandwich which in Figure 18 consists of *n-type* wafer semiconductors with free electrons. One 'slice of the bread' is an emitter where negative charge enters (opposite to the arrows) and the other is a collector where negative charge leaves. The equivalent of the common anode is the 'filling' of the sandwich and is called the base. It is a *p-type* semiconductor with positively charged 'holes'. The base circuit plays a role rather like the grid of a triode, as we saw in Chapter 4, with a low positive voltage source of around 0.5 volts between the base and the emitter. It drives the input or base current that is to be amplified. The collector circuit contains a power supply that creates a positive voltage difference between the collector and the emitter of, for example, around 9 volts.

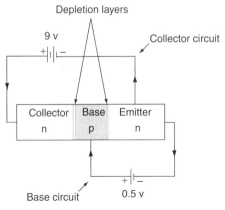

18. Transistor circuits

Let's first look at what happens in the base circuit. The *n-type* layers have free electrons and the *p-type* layer has free 'holes'. So even when no voltage is applied across the junction between the layers, some of the free electrons in the *n-type* layer flow to fill some of the holes in the *p-type* layer of the base. This continues until there is equilibrium. At this point, an insulating or depletion layer has formed between the base and the emitter. The difference in potential across the layer inhibits further movements and the transistor is totally switched off. When a positive voltage (shown as 0.5 volts in Figure 18) is applied between the base and the emitter, then the thickness of the depletion layer begins to reduce. Up to a certain level of voltage (about 0.3 volts for silicon), the voltage is not big enough to move the electrons and so no charge flows. As the voltage is increased, so more electrons begin to move creating more flow until at about 0.7 volts the depletion zone disappears and the transistor is fully switched on. In other words, as the voltage in the base circuit varies so the flow of charge from emitter to base also varies in proportion. When the electrons arrive in the base, some find positively charged holes and drop into them, but others flow around the base circuit to create a base circuit current, as shown in Figure 18.

Now we need to consider the collector circuit shown in Figure 18. The flow of charge in this circuit will be affected by the flow in the base circuit because the base is very thin. Some electrons in the base circuit get attracted towards the depletion layer between the base and the collector. The positive voltage on the *n* side of the collector circuit means that electrons will not flow easily from collector to base, but electrons in the base find it very easy to move into the collector. This means that when electrons from the emitter move into the base, some are swept across the depletion zone and into the collector and become part of the collector current. In other words, the collector current depends on the number of electrons coming from the emitter to the base – but this, in turn, depends on the thickness of the depletion layer between the emitter and base which is controlled by the voltage of the base circuit. In effect, therefore, the collector current is controlled by the base voltage. Since the base current also depends on the base voltage, then the collector current is effectively an amplified version of the base current. The size of the amplification depends on the proportion of electrons that complete the journey across the base into the collector. The ratio of input current to output current can be as little as 10 for a high-power transistor as used in a Hi-Fi, but as much as 1,000 for a low-power transistor in a hearing aid.

By the late 1950s, these electronic components were becoming very complex, and reliability needed to be improved and costs reduced. The answer came in the form of the miniaturized integrated circuit. The impact was profound. Electronic circuits were being made on a wafer made of pure semiconductor material such as silicon using photographic and chemical processing in highly specialized facilities. A few hundred transistors could be formed on one silicon chip and linked together. In the 1970s, the number of transistors integrated onto a silicon chip doubled every couple of years – a phenomenon that came to be known as Moore's Law, after Gordon Moore who co-founded Intel Corporation in the USA. Integrated circuits are found now in almost all electronic equipment. In 2009, Intel unveiled a prototype single-chip cloud computer (SCC) that has 1.3 billion transistors on a chunk of silicon the size of a postage stamp.

One way of thinking about information is to see it as being expressed in patterns – whether of letters on a page, daubs of paint in a picture, shapes carved into a sculpture, holes punched in cards, sounds in a piece of music, or connections between neurons in the brain. In digital electronics, as we will see in a moment, information is captured by using patterns of 'bits' of high and low voltage. Herman Hollerith spotted the potential for manipulating patterns of information when, in the 1880s, he used punched holes in cards to represent data read by machines. His system reportedly saved $5million for the 1890 USA National

Census. The machines were eventually electrified and big companies began to input, store, and process more data. Hollerith's company merged with three others to form what was to become IBM in 1924.

Modern computing began in the late 1930s, and it was entirely natural to use punched cards to input programs and data. The first commercial computer was the UNIVAC in the 1940s. By the early 1950s, many large companies were using computers for billing, payroll management, cost accounting, and were exploring more complex tasks like sales forecasting, factory scheduling, and inventory management. By 1960, there were a number of companies supplying large 'mainframe' computers. These were institutional machines run by specialists. Individual programmers had little access but could be given 'driving lessons'. Software for large numerical calculations enabled engineers to begin to use them, and new techniques of numerical analysis began to be developed. These first computers were physically large, as pointed out earlier. However, in the 1960s and 1970s, transistors and then microprocessors with integrated circuits took over, leading eventually to the personal computer, laptop, and hand-held media player and personal digital assistant.

Almost all modern electronics is digital – but what does that mean? At root, digital systems are based on devices that at any given moment are in one of two states – low or high voltage. The low-voltage state is near to zero. The high-volt state is at a level that depends on the supply voltage being used in a particular application. These binary levels – called bits – are often represented as (low and high) or (false and true) or (0 and 1). The big advantage is that it is easier to switch a device into one of two known states than to try to reproduce a continuous range of values. Information is captured as lists of bits.

One of the very basic devices for manipulating information is a logic gate. The simplest have two inputs (either of which could be

0 or 1) and one output. The relationship between the input and output is set out in a truth table. The most common are AND and OR. An AND gate has a truth table in which the output is true (1) only when both inputs are true (1). An OR gate has a truth table that shows the output is true (1) when either input is true (1). A NOT gate shows an output which is always the opposite of a single input. Other common gates are NAND (which is NOT AND), NOR (NOT OR), and XOR (exclusive OR, i.e. output is true if only one input, not both, is true). There are a number of different ways of making electronic devices that model truth tables, and typically they include transistors. For example, a single-transistor circuit is a NOT gate – so when the input is 0, then the output is 1, and vice versa. Transistors are connected together in different ways in different circuits designed to produce the output required for a particular logic gate. In effect, the behaviour of a logic gate 'emerges' from the way the transistors are connected.

Flip-flops are made from interconnected logic gates. They are used in microprocessors for memory storage, counting, and many other applications. A flip-flop retains a bit until instructed to forget it. So the value of its output depends on its existing bit state and an input. There are many types of flip-flop, but one is the toggle flip-flop. Its input is fed by a stream of clock pulses. Each pulse consists of a rising 'edge' (a step up from 0 to 1) followed by a falling 'edge' (a step down from 1 to 0) which causes the output of the flip-flop to change.

Flip-flops are used to count in binary numbers (see note in the references section for this chapter to remind yourself of how we count in decimals). Imagine we want to count up to 15 pulses – we will need 4 flip-flops which we will call A, B, C, D. At the start, all flip-flops are in state 0, i.e. the string of binary states in ABCD is 0000. The pulses arriving at A will comprise a rising edge (from 0 to 1) and a falling edge (from 1 to 0). So let's start by focusing on the behaviour of flip-flop A. When a pulse is applied to A, its rising edge has no effect, but its falling edge causes A to change from 0 to 1 – so one pulse will make one change. Thus, when the first pulse arrives, A switches from 0 to 1. When the second pulse arrives, A switches back from 1 to 0. A continues to oscillate between 0 and 1 at every pulse.

Now let's look at flip-flop B. It behaves as A but with its input coming from the output of A – so it only changes when the output of A falls from 1 to 0. Therefore, B also oscillates between 0 and 1 but changes only every two clock pulses. It starts as 0 and remains as 0 until the second pulse, when it switches to 1. B remains as 1 until the fourth pulse, when it changes back to 0. In a similar manner, flip-flop C changes every four pulses and flip-flop D changes every eight pulses.

What does this mean? The value in A indicates the size of 2^0 (which is always 1 – compare this with 10^0 in a decimal count). In other words, when it is 0 we know there has been an even number of counts (half the number of pulses) and when it is 1, we know we have an odd number. The value of the flip-flop B indicates the size of 2^1 (which is always 2 – compare with 10^1 which is always 10 in a decimal count). So when B is 1, we have a decimal number containing a count of 2, and when it is 0, then we don't. Likewise, flip-flop C shows the size of 2^2 (which is 4 – compare with $10^2 = 100$ in a decimal count) and flip-flop D shows the size of 2^3 (which is 8 – compare with $10^3 = 1000$ in decimal). Thus, for example, if the flip-flop sequence from D to A contains 1011, then the count has been $(1 \times 2^3) + (0 \times 2^2) + (1 \times 2^1) + (1 \times 2^0) = 8 + 0 + 2 + 1 = 11$ in decimal. The maximum number we can get is 1111 = 15 in decimal, so by using 0000 as well, we can count up to 16. Again, we can see just how a behaviour, in this case counting, 'emerges' from the interconnections between the parts – the flip-flops.

Obviously, in a practical counter we need more than 4 bits. 8 bits are called a byte, enabling us to count to 256 in decimal – again by using 0 through to 11111111 (i.e. 0 to 255 in decimal). But even with 16 or 32 bits, we are restricted. Floating point numbers are a clever way of getting around the restriction – for example, a very large decimal number such as + 53,678,900,000,000,000,000 becomes + 5.36789×10^{19}, i.e. a number (called the significand or mantissa) with only one figure before the decimal point but others after it, multiplied by 10 to the power 19 (called the exponent). We can turn that into a binary floating point number, so that + 5.36789×10^{19} becomes the bit pattern shown in Figure 19 where s is the sign (0 for plus and 1 for minus).

But how does the computer add and subtract numbers? How does it hold and store symbols? Just as integrated circuits are manufactured as logic gates and flip-flops, so they are also made to do arithmetical operations such as adding, subtracting, and multiplying. Characters and symbols are each represented by a 7-bit code. The ASCII (American Standard Code for Information Interchange) defines the standard way this is done. So, for example, a capital A is 1000001, capital B is an A plus 1, and so is 1000010. Capital Z is 1011010. Whenever you use a word processor, deep inside the integrated circuits of your computer are all of the letters you have typed in but memorized and manipulated as lists of bits. All these bits are passed through tiny integrated circuits as they are processed for a particular function such as a spell-checker.

So at the heart of any digital system, whether mobile phone or computer, is a stream of bits. In a digital radio (remember your mobile phone is a radio), the carrier wave we referred to in Chapter 4 is modulated (altered or varied) by a stream of bits. Pulse-code modulation is used for converting continuous analogue signals into digital ones. The continuous signal is

0	1	1	0	0	0	0	0	0	0	1	1	1	0	1	0	0	0	1	1	1	1	0	0	0	1	1	1	0	0	0	1
s		exponent								mantissa or significand																					

19. A 32-bit pattern

sampled at intervals and the amplitude is translated into a stream of numbers that are sent out as bits. The bits and the bytes can then be manipulated for many purposes, such as inverting an image or cutting and pasting words and sentences. Examples are many and various in radios, televisions, cameras, video and music, multiple sensors used in arrays such as radar and CCTV, statistical data, communication devices, the internet, biomedical applications such as hearing aids or heart pacemakers, engineering data analysis such as recordings of earthquakes or seismic events. Data can also be more easily encrypted to keep information secure. Engineering designers can do so much with the bit patterns of digital electronics. For example, computer software is used in cardiovascular mechanics research to study the highly nonlinear response of tissues/organs to investigate microcalcification on the vulnerable plaque mechanics in the arteries, as vulnerable plaque rupture is one of the leading causes for sudden heart attacks.

By the late 1960s, the average US company was devoting less than 10% of its capital equipment budget to information technology (IT); 30 years later, it was 45%; and by 2000, it was the size of all other equipment combined. In those 30 years, businesses changed and computers changed. Businesses became more service-oriented and computers became smaller, cheaper, easier to program, and more powerful, with a big increase in the range of tasks to which they could be applied. Personal computers were soon being wired together in networks to allow the exchange of files and share printers. The old mainframes didn't disappear – they were transformed into new kinds of data centres – the engine rooms of modern businesses.

In the 1980s, optical fibres began to be used to transmit telecommunications data over long distances. They are glass or plastic fibres that carry light along their length faster than other forms of communications. They are better than metal wires because they offer higher bandwidth, are lighter in weight, have low transmission loss, are less sensitive to electromagnetic interference, and are made from cheap material. Nicholas Carr observes that the fibre-optical internet is doing for computing what the AC network did for power distribution because where the equipment is located is not important to the user. The lone PC is giving way to a new age called cloud computing in which computing is a utility just as is electricity. Computing is sold on flexibly on demand – users can buy as much or as little as they want. Interactive websites don't just provide information – they facilitate interaction between users, as in social networking sites. The social impact of these technologies will soon be as great as the steam railway or the automobile.

So given that the technology has been driven by dramatic miniaturization, how small can we get in the future? Quantum computing is a fast-developing research topic in which bits of 0 or 1 in electronic computing are replaced by quantum phenomena at the atomic level. This technology may provide as big a change in scale as that from valves to transistors. Nanotechnology works with very small particles that are comparable to the size of a biological cell. A nanometre is one billionth of a metre, or one millionth of a millimetre. A cell measure 10 to 100 nanometres, and a gene may be 2 nanometres wide with a length similar to a cell. Magnetic nanoparticles offer some attractive possibilities in biomedicine. For example, they can be made to bind to a cell giving that cell a tag or 'address' – in a somewhat similar manner to a computer memory address. They can then be used for imaging, tracking, or as carriers. As they are magnetic, they can be manipulated by an external magnetic field. They can be made to deliver a package, such as an anticancer drug to a tumour. The particles can be made to resonate to heat up and hence kill tumours or act as targeted agents of chemotherapy and radiotherapy. The research could lead to better tools for screening different diseases in a non-invasive and accurate way and for administering therapeutic agents safely and effectively with fewer side effects.

Another technique called photodynamic therapy uses light to destroy cancer cells. The patient is injected with a photosensitizer which accumulates in a tumour. Using a precise laser, the tumour is then targeted with light of a wavelength that will be absorbed, producing a drug that kills the cell. The big breakthrough will be to translate innovations in the laboratory into commercially viable medical products.

The IT revolution is changing into an age of systems as we begin to understand how complexity can emerge from simplicity. Computers certainly display this through the three general characteristics of systems thinking – layers, connections, and processes.

Chapter 6
The age of systems – risky futures

'We must ensure that this never happens again.' How often we hear these words after an inquiry into a failure. But can they be delivered? With the exceptions of Chernobyl and Fukushima, in earlier chapters we have concerned ourselves only with engineering achievements. Major disasters are fortunately quite rare, but when they do happen they hit the headlines because of large-scale damage and number of people killed.

In this chapter, we will contemplate the inevitable gaps between what we know, what we do, and why things go wrong. As we explore modern ideas of systems complexity, we will see that the gaps are filled by risk. Risk is at the heart of major engineering questions. How do we know what is safe? How safe is safe enough? How do we ensure that the London Eye won't fall over? What are the chances of another Chernobyl or Fukushima? Are engineering failures really failures of engineers? Why do so many major projects seem to be over budget? How do we learn from failure so history does not keep repeating itself?

We know from everyday life that things don't always turn out as we want. This is also true of engineering – indeed, making decisions in everyday life has much more in common with engineering practice than may seem at first sight. Both require us to use common sense in solving problems. We have to decide what we want, what we think we know, how we may achieve what we want, what actually to do, and finally what we think might be the consequences. We know if things don't work out as we hoped, then we will be affected in all sorts of ways – varying from minor upset to deep and major harm. Engineering decisions, however, affect many more people than do our everyday ones and may expose them to all kinds of risks – including death. So, quite rightly, what engineers do is closely scrutinized and can, ultimately, be tested in a court of law as a duty of care.

Engineering is practical – it is about creating tools that work properly, which simply means that they are fit for their intended purpose. Just as our everyday decisions are constrained by financial, social, political, and cultural situations, so are those made by engineers. Perhaps the most obvious constraint is finance because almost all engineering is a business activity. Engineering activity must be affordable, but the predictability of final cost is almost as important as the amount. Politics is an example of a less obvious constraint. For example, natural hazards such as earthquakes and wind storms happen all over the world – but when they occur in regions where buildings have not been properly engineered, the consequences are much more devastating than they should be. The large-scale loss of life that happened in Haiti in 2010 could have been avoided if the right technology had been in place – but the reason it wasn't is a political issue.

Failure is a very human condition – none of us like it, but most of us have to deal with it. It is a misunderstanding if one thinks that successful people have never failed – what makes them successful is the way they cope with failure, learn from it, and try never to repeat it. One of the first engineering failures that led directly to improvement happened about 5,000 years ago, according to Egyptologist and physicist Kurt Mendelssohn. The Egyptians wanted to develop the first stepped pyramids into the more familiar smooth-sided ones. The Bent Pyramid at Dashur was being built when the Meidum Pyramid collapsed. The builders realized that the slope was too steep and so reduced it, turning the pyramid into the now famous 'bent' shape. Man-made disasters within living memory vary from large-scale explosions at Flixborough, UK (1974) and Bhopal, India (1984), to the major fire that destroyed the Piper Alpha offshore oil rig, UK (1988). In 2010, a major oil spill occurred in the Gulf of Mexico after a BP oil rig exploded – one of the biggest environmental disasters in US history.

In Chapter 1, we referred to the disturbing cloud around the mystery, complexity, and opacity of modern engineering tools. Part of that cloud is a sense of helplessness when our computers sometimes just decide to behave in ways we don't want, or central heating gas boilers break down just when we need them during a cold snap. We have an uncomfortable concern that whilst the power of technology is bringing benefits, it is also bringing a sense of alienation and a feeling of vulnerability with more potential for doing harm. Many of us question whether we have come to rely on engineering and technology almost too much. We wonder if the risks are beginning to outweigh the benefits.

Of course, we cannot undo the past – but we can do things differently in the future, and our understanding of and attitude towards risk has a key part to play. We have to accept that science, technology, engineering, and mathematics cannot deliver certainty nor guarantee a future with no failures. But what can be delivered is an acceptable level of risk which is comparable with those from natural events. The problem is that the definition of what is an acceptable risk is not straightforward. We all know that people are killed in car accidents but that does not stop us from driving – most of us consider the balance between risk and benefit to be acceptable. Some people hate to fly – they see the risk/benefit as unacceptable. Yet statistics show that flying is much safer than driving. Our perceptions of risk vary greatly in complicated ways. One of the main deciding factors is the degree to which the risk is familiar and we feel in control. Our attitude to failure is not simple either. A specific failure event, such as a plane crash or factory fire, may well make us more aware of a particular type of risk. But the event itself does not necessarily make that risk unacceptable, as some are interpreting the events at Fukushima in 2011. Of course, it is possible that it is unacceptable – but the fact of failure does not make it necessarily so.

The simple reality is that what we think we know is always incomplete – there is always a gap between what we think we know and what we actually do know – so when we act, there is bound to be a gap between what we do and what may happen. These gaps are important, but often forgotten and neglected when we revel in our successes. Incompleteness is that which we do not know. Plato wrote, 'How can we know what we do not know?' That is a fundamental challenge to the management of risk and uncertainty, and it is one that should make us always work with a degree of humility. Indeed, the incompleteness gap was famously ridiculed in the media when Donald Rumsfeld, the one-time US Secretary of State for Defense, said:

> There are known knowns – these are things we know we know. There are known unknowns.…these are the things we know we do not know. But there are also unknown unknowns; these are the things we don't know we don't know.

But Rumsfeld was right. We can see this if we examine how we learn. As a small child, there are many things that you didn't know existed – like Newton's laws. They may be known by others but not by you – you are completely unaware – to you, it was an unknown, unknown – perhaps it still is. Those of us who went on to study physics eventually learn about Newton's laws, and so for these people Newton's laws become a known known. But perhaps, like most people, they stop studying physics before getting to Einstein's relativity theory. They know it exists but that's all – it is a known unknown. We all have lots of these. But the really interesting examples of incompleteness are the unknown unknowns, where not just you as an individual but no one anywhere knows. The failure of the Dee Bridge in 1847 is an historical example where the actual mechanism of failure, called lateral torsional buckling, was only understood some 50 years after the event.

How do engineers actually deal with risk? They do their utmost to make it acceptably small in two distinct ways. The first is to make sure the physical tool works properly with a good safety margin. The second is to make sure that people and organizations are well managed so that the risk of error is controlled. Traditionally, these two ways of working are seen as quite distinct – one 'hard' and objective, and the other 'soft' and subjective.

The simplest way of minimizing the risk of failure of a physical 'hard' tool is to use a large safety factor. This is the ratio of the capacity of a engineered tool to the maximum demand imposed on it. For example, the safety factor will range from 1.1 to 2 for the structure of a building. This means that the estimated strength of a particular beam (capacity) in the building may be up to twice that of the biggest expected weight to be placed on it (demand). Some engineers will calculate statistical estimates of the chance that the demand might exceed the capacity – this is called reliability engineering but is fraught with difficulties.

Safety factors on their own aren't enough. Engineers know that no matter what they calculate equipment does fail, humans make mistakes, and natural hazards, such as earthquakes, do occur. So they often have back-up, or contingency, plans – this is known as defence-in-depth. The basic idea is to try to prevent an accident in the first place with appropriate safety factors but then to limit the progress and consequences if one should occur. The engineering team looks at all of the possible demands they can think of, and they try to make the chance of failure acceptably small. For example, they want to make sure your car will always start when you want it to – so they look at the reasons why your car might not start. In a similar way, engineers examine the safety of a nuclear reactor by drawing enormous logic diagrams covering many pages which trace how an event (such as a pump that fails to circulate cooling water in the nuclear reactor) might affect other parts of the system. These are called event trees. They also draw diagrams that show how a fault may have been caused by other credible faults – these are called fault trees. A partial fault tree for your car not starting is shown in Figure 20. Clearly, all possible faults are not equally likely, so engineers will assess the relative frequencies of faults. For example, the most likely reason why a car won't start is electrical – a flat battery – so that is the first thing anyone looks for. The least likely is a broken crank shaft and you might find that only after extensive examination. The purpose of the fault and event tree analysis is to enable the engineers to understand the interactions between the various components of a system, and hence to design possible defence-in-depth schemes if appropriate, and therefore reduce the chances of all failures of components (such as a battery) to a level where the total chance of failure is acceptably small.

But some tools have become so very complex that it is now impossible to draw an event tree for all possibilities. Engineers have therefore started to work and manage the risks in layers – as we began to outline for electronic equipment in Chapter 5. A computer is a good example. We know that deep inside a computer is a layer of interconnected semiconductors consisting of the transistors and other components that work together to make the logic gates in the next layer up. The gates work together in a third layer to make flip-flops and other devices making up patterns of bits and bytes. Next are the registers, memory cells, and arithmetic units made of connected flip-flops and all put together into integrated circuits that make up the brains of a computer – the microprocessor or central processing unit (CPU). The top hardware layer contains several other connected components such as memory, disc drives, monitor, power supply,

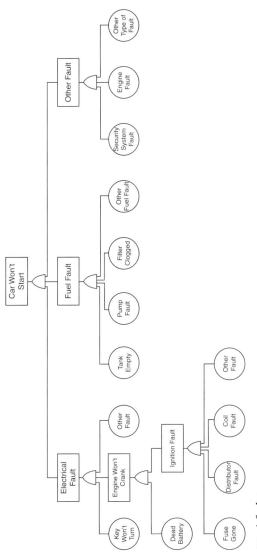

20. **A fault tree**

keyboard, and mouse, all working together to make up the characteristics of a particular machine. The next layer is the first software layer. Software is the set of instructions that make the microprocessor work. The lowest of these are the machine languages in which instructions to the binary system are programmed. Only specialists can work in that language. To make it possible to write general computer programmes, the next layer contains the programmable languages such as Fortran, Algol, and C. These are easier to understand and use, and are translated into machine language by the computer so they can be processed as patterns of bits. Then we have the operating systems such as Windows. Finally are applications such as word processors, spreadsheets, and other more specialized programs, such as finite element packages to help structural engineers calculate the internal forces in a bridge or to simulate processes in a chemical plant. Very few people can understand how the system works at all levels, so engineers have to specialize to cope with the detail and manage the risks at a particular level and leave the other levels to other specialists. So a hardware specialist in transistors will be different from one specializing in digital systems and different again from a software specialist. These specialisms pose new risks since the relationships between levels and, more importantly, between the ways of understanding of specialists in those levels, are not straightforward.

The story of earlier chapters tells how the growth and success of engineering and technology has largely been due to our increased understanding of physical phenomena. But what has not advanced so quickly is our understanding of ourselves – how we organize to achieve the things we want. The IT revolution of the 20th century, together with advances in biochemistry and our understanding of the chemistry of DNA, are perhaps the ultimate expression of the success of reductionist science. But now, in the 21st century, we are beginning to understand how complex behaviour can emerge from interactions between many simpler highly interconnected processes. We are entering into the age of systems with a potential for new risks through interdependencies we may not fully understand. For example, we now know that some (but not all) physical processes are chaotic, in the sense that, whilst they may appear to be reasonably simple, they are inherently difficult to predict. We have discovered that they may be very sensitive to very small differences in initial conditions and may contain points of instability where paths diverge. Consequently, two identical processes that start with almost but not exactly the same initial conditions may diverge considerably after only quite a short time. We see this even in quite simple systems like a double pendulum as well as bigger and more complex systems like weather forecasting. This is a new kind of uncertainty that presents a new kind of risk. Highly interconnected systems, such as electrical power supply networks, the internet, traffic highways, and even building structures, can become vulnerable to quite small damage cascading to disproportionately large consequences. Even if the chance of the initial damage is very low, the consequences can be very severe. Such systems lack resilience or robustness. We have to learn to live with the knowledge that we cannot predict the total behaviour of a complex system from the performance of its interdependent parts – we have to expect the unexpected unintended consequences.

As a result, some engineers have begun to think differently – they use what many call 'systems thinking'. A system is a combination of things that form a whole. Immediately, we can distinguish a 'hard system' of physical, material objects, such as a bridge or a computer, from a 'soft system' involving people. Hard systems are the subject of traditional physical science. They comprise objects as tools that all have a life cycle – they are conceived, designed, made, used, and eventually discarded, destroyed, or recycled. They have external work done on them, and they respond by doing internal work as they perform. In effect, they are 'manipulators' of energy – processes of change in which energy is stored, changed, and dissipated in specific ways depending how the elements of the systems are connected. A complex hard system has a layered structure just as in a computer. At any given level, there is a layer underneath that is an interconnected set of subsystems, each of which is a hard system and also a physical process. This kind of hard system thinking enables us to see commonalities between different specialisms that were previously thought to be different. Each object in each layer is a process driven by a difference of potential – an 'effort' to cause a flow which is opposed by impedance. For example, the difference in height between two ends of a water pipe causes the water to flow from the high to low end. The difference in the change of velocity of a mass causes the flow of internal forces. A difference in temperature between two points in a body causes heat to flow from the hot to the cold. A difference in the voltage across two electrical terminals causes a flow of electrical current measured in amps.

Processes interact with each other – we can think of them as sending 'messages' about their own 'state of affairs' to their 'friends and neighbour' processes. The power of a process is the input effort times the resulting flow, e.g. watts are volts times amps. Power is also the rate at which energy is used. In other words, energy is the capacity for work in a process and is an accumulation of power over time. During these processes, some of the impedance dissipates energy (resistance), some of it stores potential (capacitance) some of it stores flow (induction). For example, the dissipation or loss of energy through a resistance in an electrical circuit is equivalent to the dissipation of internal energy due to the damping of the vibrations of a child's swing or the vibrations of a bridge. The storage of energy of the charge in a capacitor is equivalent to the storage of the kinetic energy of velocity through the rotating mass or inertia of a flywheel. The storage of the energy in an inductor in an electrical circuit is equivalent to the strain energy stored due to the flexibility of a bridge structure.

Every hard system is embedded in a soft system – ultimately all of us. The physical world 'out there,' outside all of us, exists. The problem is that we can only perceive it through our senses. We can create our own mental models and discuss them between us in a way that enables us to come to some agreements. We can then act in ways that seem appropriate. An engineering team is an example of a specific soft system. It has the job of understanding a particular hard system, predicting how it may behave in various situations, and ensuring that it can be controlled safely. The team must also try to envisage how their hard system may be used by other soft systems – the users – us. Soft systems are the subjects of the social and management sciences.

A crucial question at the heart of all soft systems is: How do we judge the quality of information on which we depend to make decisions that could risk someone dying? Like the rest of us, engineers want information that they know to be true. If information is true, then we can use it without concern. But what is truth? Philosophers have been discussing this since Plato. In engineering, as in everyday life, we need a practical commonsense view that helps us to manage acceptable risks. So we accept that a true statement is one that 'corresponds with the facts'. But what are facts? Facts are self-evident obvious truths. We have an infinite regress since facts are true statements – we have defined something in terms of itself. In everyday life, for most of the time, this just doesn't matter. In engineering, because of the duty of care for people's lives, we must examine the notion of truth and risk a bit further.

In pre-modern society, there were broadly two ways of arriving at truth, *mythos* and *logos*. *Mythos* derived from story-telling. It was often mystical, religious, emotional, and rooted in the subconscious mind. It required faith – belief that cannot be proved to the satisfaction of everyone else – and lacked rational proof. *Logos*, on the other hand, was rational and pragmatic, and was about facts and external realities – the kind of reasoning we use to get something done. Church-building and church art was a physical expression of faith – the expression of an emotional truth of *mythos* through the rational tools of *logos*. To our modern minds, engineering and science seem to spring only from practice based on *logos*, but the distinction between *logos* and *mythos* has never been, and perhaps never will be, one of total clarity. Spiritual faith springing from *mythos* is genuine and real to the believer and an important basis for the way we live our lives because it provides answers to deep questions about the meaning and purpose of life – but it isn't testable in an objective way. Knowledge in STEM is shared objective information that is outside of any one individual. Unfortunately, there are two big problems. First, it has become highly specialized and quite fragmented so that the newest details are understood only by a relatively small number of specialists. Second, it is never simply factual and total, but is always incomplete, as we discussed earlier through the remarks of Donald Rumsfeld.

So where does this leave us? As we design newer and faster computers, and as we discover more about how our brains work, the more we will be able to do. For example, we will build intelligent robots that can carry out a whole variety of tasks ranging from self-controlled vehicles to body implants. Such developments depend on the modern view that knowledge and information is sets of layered patterns in our computers and in our brains. Of course, we understand how patterns of bits in our computers work, but our understanding of the brain is still developing. Nevertheless, since these patterns represent something other than themselves, they can only be a depiction, or a model, of the world around us. Whatever the source of our beliefs, spiritual or practical, *mythos* or *logos*, what we do is based on what we believe to be true – in other words, on what we think we know. The incompleteness gaps between what we know and what we do and what might be the consequences – intended or unintended – are filled by faith. In *mythos*, this can lead to major tensions between religious sects.

In the *logos* of STEM, faith is a small but inevitable aspect of risk that engineers seek to minimize but can never eliminate. These gaps are often denied, ignored, or misunderstood but are actually of critical importance in risk.

The job of the engineer is to make the risks acceptably small. In doing so, engineers do not look for truth – that is the purpose of science – rather, they look for reliable, dependable information on which to build and test their models of understanding. They are acutely aware of context. For example, they know that Newton's laws of motion only apply when objects are not travelling close to the speed of light. So whilst these laws are not strictly true, they are dependable for most engineering systems. The ultimate test is that the engineer who uses them is acting with a duty of care as tested ultimately in a court of law. So the commonsense idea of truth as 'correspondence to the facts' (unaware of the philosophical infinite regress) is also the engineering sense of dependability. Ultimately, it relies on the final sanction of a duty of care in a law court.

No matter how we judge the quality of information, it is clear that layers and connections are very important in the way we understand complex systems and how we are able to demonstrate a proper duty of care. Arthur Koestler coined a very useful word for thinking about layers. He suggested that the word 'holon' should refer to something, indeed anything, which is both a whole and a part. So a logic gate is a whole in the sense it has an identifiable function such as AND or OR that emerges from the working together or connectedness of its parts – the subsystems of transistors and other components that make it function as required. But the logic gate is also a part in the way it works with other components to form a flip-flop with an emergent function, such as binary counting. From this, we can conclude something very interesting. The characteristics of each layer emerge from the interacting behaviour of the components working in the layer below.

Emergent properties of holons are also found in soft systems. Indeed, you and I are examples through our ability to walk and talk. None of your parts can walk or talk on their own. The net result is that you are more than the sum of your parts. You are the result of the active interaction of your parts with your environment in a process we call living. This same argument works at every level. Looking inwards, your structural subsystem of bones and muscles is also a holon with its own emergent properties such as your body size or muscular dexterity. Looking outwards, your family is a holon with its own emergent properties such as happiness or closeness. The highly connected neural connections in the brain create emergent consciousness. Well-maintained connections between people make for good relationships. Well-maintained connections between physical elements make for good physical systems such as bridges. Just as the complexities of well-engineered hard systems have reached the limits of our ability to understand the interdependencies between components, so we have begun to recognize the close interdependencies between hard systems within soft systems. But soft systems are also hard systems, since flesh and blood consist of atoms and molecules – the complication is that we have multiple layers of human intentionality which we simply cannot model. Put at its simplest, intentionality is having a purpose, aim, or goal. It is this multiple-layered interacting intentionality that makes soft systems so difficult.

The potential that drives the flow of change in a soft-system process stems from a need or a want – the answers to our questions *why*. Peter Senge called it a creative tension between 'where we are now' and 'where we want to be in the future'. The flow of change is captured by asking questions and tracking the answers about *who*, *what*, *where*, and *when*. For example, *who* questions may be about the effects of changes in personnel occupying key roles. *What* questions concern choices, measurements, and monitoring of performance indicators including evidence of potential problems and/or success. *Where* questions are issues of context and the impact of changes in context; and *when* questions are about timing. *How* questions are about policies, methodologies, and procedures – they are the way the change parameters of *who*, *what*, *where*, and *when* are transformed. One way of envisaging the relationship between these factors is *why = how (who, what, where, when)*. This is not a mathematical formula but is intended to capture the idea that a process is driven by the potential difference of *why* creating a flow of change in *who*, *what*, *where*, *when* through transformations *how*. We can speculate that impedance in soft processes is made up of factors analogous to hard systems. So resistance is a loss of energy perhaps due to ambiguity and conflict. Capacitance is an accumulation of our ability to do things or to perform. Inductance is our capacity to adapt and innovate. This way of 'systems thinking' is beginning to provide a common language for hard and soft systems, though there is still some way to go to make it totally effective and many engineers have not yet embraced it.

How do we deal with risk in this complexity of multiple layers of emergent characteristics? We take a tip from the medics. Forensic pathology is the science or the study of the origin, nature, and course of diseases. We can draw a direct analogy between medics monitoring the symptoms of ill health or disease in people with engineers monitoring the symptoms of poor performance or proneness to failure in an engineering system such as a railway network. Structural health monitoring is one existing example where measurements of technical performance of a complex piece of equipment are made to detect changes that might indicate damage and potential harm before it becomes obvious and dangerous. At present, its focus is purely technical with axioms that include the assertion that 'all materials have inherent flaws or defects'. This helps us to see that less than perfect conditions may have existed and gone unrecognized in a hard system for some time – there is potential for damage to grow that perhaps no-one has spotted. Under certain circumstances, those conditions may worsen and it is the duty of those responsible to spot the changes before they get too serious. We can think of these hidden potential threats as hazards, i.e. circumstances with a potential for doing harm, or 'banana skins' on which the system might slip. We attempt to detect these hazards by looking for changes in important measurements of performance. For example, steam railway wheel tappers used to check the integrity of steel wheels by striking them with a hammer – a change in the sound told them that the wheel was cracked. Hazard and operability studies (called Hazop) are widely used in designing chemical engineering processes to identify and manage hazards. The safety of a hard system may also be protected by controlling the functional performance of a process automatically. Engineers design into their hard systems feedback tools that operate on the inputs to make the desired outputs. Watt's centrifugal governor (Chapter 3) to control the speed of a steam engine by changing the input flow of steam was an example. Water supply (see Chapter 1), speed controllers on cars, aircraft landing systems, and space craft are amongst the many examples where control engineering is now used.

Just as there are technical hazards in hard systems so there are human and social hazards in soft systems. For a soft system, the axiom noted earlier becomes 'all soft systems have inherent flaws or defects'. Social scientists such as Barry Turner, Nick Pidgeon, Charles Perrow, and Jim Reason have studied many failures including those mentioned at the start of this chapter. They have discovered that human factors in failure are not just a matter of individuals making slips, lapses, or mistakes, but are also the result of organizational and cultural situations which are not easy to identify in advance or at the time. Indeed, they may only become apparent in hindsight.

For example, Jim Reason proposed a 'Swiss cheese' model. He represents the various barriers that keep a system from failing, such as good, safe technical design, alarms, automatic shutdowns, checking and monitoring systems, as separate pieces of cheese with various holes that are the hazards. The holes are dynamic in the sense that they move around as they are created and destroyed though time. The problems arise if and when the holes in the 'cheese' barriers suddenly 'line up.' In effect, they can then be penetrated by a single rod representing a path to failure – a failure scenario. The hazard holes in the cheese arise for two reasons. First are the active failures – unsafe acts by individuals are an example. Second are latent 'pathogens' already resident in the 'cheesy' system. These pathogens result from various decisions and actions that may not have immediate safety consequences but which may translate into later errors – for example, where undue time pressures result in some 'corners being cut' in order to reach a time deadline on a particular project. The pathogens may lie dormant and unrecognized for many years – they may only be discovered when the cheese holes line up.

Barry Turner argued that failures incubate. I have described his ideas using an analogy with an inflated balloon where the pressure of the air in the balloon represents the 'proneness to failure' of a system. The start of the process is when air is first blown into the balloon – when the first preconditions for the accident are established. The balloon grows in size and so does the 'proneness to failure' as unfortunate events develop and accumulate. If they are noticed, then the size of the balloon can be reduced by letting air out – in other words, those responsible remove some of the predisposing events and reduce the proneness to failure. However, if they go unnoticed or are not acted on, then the pressure of events builds up until the balloon is very stretched indeed. At this point, only a small trigger event, such as a pin or lighted match, is needed to release the energy pent up in the system. The trigger is often identified as the cause of the accident but it isn't. The over-stretched balloon represents an accident waiting to happen. In order to prevent failure, we need to be able to recognize the preconditions – to recognize the development of the pressure in the balloon. Indeed, if you prick a balloon before you blow it up, it will leak not burst. Everyone involved has a responsibility to look for evidence of the building pressure in the balloon – to spot the accident waiting to happen – and to diagnose the necessary actions to manage the problems away.

The problems we are facing in the 21st century require all of us, including engineers, to think in new ways. The challenge for STEM is to protect the important specialisms that allow us to progress our detailed work whilst at the same time providing a set of integrating ideas that allow us to see the big picture, to see the whole as well as the parts – to be holistic, but to keep the benefits of reductionist science. Engineering systems thinkers see the many interacting cycles or spirals of change that we have identified in earlier chapters as evolutionary developments in knowledge (science) and action (engineering) that leapfrog over each other. But it is not the Darwinian evolution of gradual accumulation – it is purposeful human imagination used to improve our quality of life, so that to act you need to know and to know you need to act. This view is in direct contrast to a reductionist philosophy that sees knowledge as more fundamental than action. Systems thinkers value knowing and doing equally. They value holism and reductionism equally. They integrate them through systems thinking to attempt to get *synergy* where a combined effect is greater than the sum of the separate effects. It is a new philosophy for engineering. It may sound a little pretentious to call it a philosophy – but it does concern the very nature of truth and action, since risk is as central an idea to systems thinking as truth is to knowledge. Put at its simplest, truth is to knowledge as the inverse of risk is to action. The intention of knowledge is to achieve understanding whereas the intention of action is to achieve an outcome. Truth is the correspondence of understanding with 'facts'. Risk is a lack of correspondence of outcome with intended consequences. So a degree of truth between true and false is analogous to a degree of risk between failure and success.

At the start of this chapter, we asked whether we can ensure that a specific failure will never happen again. How do we know what is safe? How safe is safe enough? The answer is that we cannot eliminate risk, but we can learn lessons and we can do better to make sure the risks are acceptable. We can punish negligence and wrongdoing. But ultimately, 'safe enough' is what we tolerate – and what we tolerate may be inconsistent because our perception of risk is not straightforward. We can ensure that structures like the London Eye won't fall over, or the risks in a railway signalling systems are acceptable, by making sure that those in charge are properly qualified and really understand what they are doing. But the chance of the London Eye falling over is not zero; the chance of a train crash due to a signalling failure is not zero. The truths of science are not enough to ensure a safe engineered future for us all – engineers use science, but there is a lot more to it than simply applying it. On top of that, all of the activity is embedded in societal processes in which everyone has a part. In a democracy, good choice requires good understanding. Informed debate should enable us to find more consistent ways of managing engineering risk.

Climate change may well be the defining test. The debate has to move on from the questions of whether change is man-made. We don't know for certain – but the evidence is overwhelmingly strong. The stakes are so high that we need to organize ourselves for major weather events which, if we are lucky, will not occur. Engineers have to deliver sustainable systems with low throughput of material and energy with more recycling. More attention needs to be given to making systems durable, repairable, adaptable, robust, and resilient.

The tensions are pulling the traditional divisions between engineering disciplines in a number of opposite directions, and they are creaking under the strain. Specialization has lead to fragmentation and a loss of overview. Engineers who hunker down in their silos cannot contribute to challenges that do not fit into traditional boxes. There is still a need for highly specialist expertise that has to keep up with new technology. But perhaps even more importantly, that specialism must be tempered with a much wider understanding of the big picture than has typically been the case in the past.

We will end our story where we began – engineering is about using tools to do work to fulfil a purpose. Over centuries, we have created some very large complex interconnected systems that are presenting new vulnerabilities, risks, and challenges. Climate change is forcing us to focus on energy – the capacity to do work. The laws of thermodynamics tell us we can shift energy around but we can't destroy it. However, as we shift it around, some of it is lost to us – irretrievably no longer available to do work – we cannot get something for nothing, entropy inexorably increases. The energy performance of buildings is a good example of the need to do better. The UK Royal Academy of Engineering says that too often no-one holds an overview and so the engineering solutions sometimes lack coherence. 'Embodied energy' is presently rarely considered – this is the energy used to make all of the materials and components to be used in a particular building before they reach a construction site. Exergy analysis is hardly ever used. Exergy is related to entropy and is a measure of the available work in a system that is not in equilibrium with its surroundings. Using it we can capture both the quality of available energy as well as the quantity. For example, it tells us that it is inefficient to use high-grade electricity from the national grid for low-grade domestic heating that takes us from ambient to around 20 degrees C.

Energy, entropy, and exergy are examples of how the challenges of the 21st century require the engineering disciplines that are much better at integrating their expertise to find synergy. If we are to make and maintain highly reliable and sustainable complex systems, then we need more of our specialist engineers to be systems thinkers that can deal both with the detail and the big picture – a synergy from the integration of reductionism and holism.

Glossary

ASCII American Standard Code for Information Interchange.

bandwidth The smallest range of frequencies within which a particular signal can be transmitted without distortion, or the speed or capacity of data transfer of an electronic system.

bit A binary digit, i.e. 0 or 1; off or on.

byte A combination of *bits* to represent a letter or a number.

capacitance Holding electrical *charge*, or more generally storing flow to create potential.

charge A basic property of matter that creates electric and magnetic forces and is positive or negative.

commutator A split ring or disc that changes the direction of an electrical current.

diode A device in which current can only pass in one direction.

emergence The way properties of a system develop out of interactions between simpler systems. Our ability to walk and talk are emergent properties since none of our subsystems (skeletal, muscle, blood circulation, nervous, etc.) can walk and talk on their own.

energy A capacity for work – potential energy is due to position, kinetic energy is due to movement, and strain energy is potential energy in a deformed material. Measured in joules.

entropy The loss of available energy in a heat engine or loss of information in a signal – an increase in disorder.

exergy The maximum useful work possible as a system moves into equilibrium with its surroundings.

field A region of space in which any object at any point is influenced by a force – gravitational, electrical, or magnetic.

galvanometer A device for detecting and measuring small electrical currents.

holon Something which is both a part and a whole at the same time.

induction A process in which electricity or magnetism produces electricity or magnetism in another body without any physical contact. More generally storing potential to create flow.

logos Rational and pragmatic reasoning about facts and external realities – the kind of reasoning we use to get something done but which says little about religion, emotions, and the meaning and purpose of life.

mythos Understanding derived from story-telling, often mystical, religious, emotional, and rooted in the subconscious mind. It requires faith – belief that cannot be proved to the satisfaction of everyone else – and lacks rational proof. Used to give meaning and purpose to life.

radiation Process in which energy is emitted as particles or waves.

reductionism The idea that a system can be completely understood by understanding its parts or components, i.e. the whole is merely the sum of its parts.

resonance Where the frequency of a stimulus is close to the natural vibration frequency causing very large vibrations.

semiconductor A material, like silicon, that conducts electricity, but not as well as a good conductor, like copper.

STEM Science, technology, engineering, and mathematics.

systems complexity Where a system has properties that are not obviously emerging from the interaction of the parts.

systems thinking A way or philosophy of approach to problem-solving that values both parts and wholes, i.e. one that combines reductionism with holism.

thermionic emission Flow of electric charge induced by heat.

transformer A device that transfers an AC from one circuit to another and changes the voltage up or down.

transistor An electronic device made from *semiconductors* that amplifies, oscillates, or switches the flow of current. Two common types are the bipolar and the field effect.

triode A *vacuum tube* with three elements, anode, cathode, and control grid, usually used to amplify a signal.

turbine A rotary engine where a continuous stream of fluid turns a shaft to drive a machine. For example, a gas turbine is driven by a flow of gas (usually air or air/fuel mixture), whereas a water wheel is driven by a flow of water.

turbofan jet A type of jet engine in which some of the air bypasses the main jet. Noise is reduced, thrust increased, and fuel consumption reduced.

vacuum tube Also known as an electron tube or thermionic valve. They resemble incandescent light bulbs and rely on thermionic emission. The two main types are *diodes* and *triodes*.

vulnerability Susceptibility to small damage causing disproportionate consequences.

winding A wire coil usually in an electromagnet. A field winding is a number of coils around individual poles connected in series, i.e. end to end so that the same current flows.

work A transfer of energy as a force moves a distance, i.e. force times distance – measured as horsepower or joules.

References

Preface

House of Commons, Innovation, Universities, Science and Skills Committee, Fourth Report, *Engineering: Turning Ideas into Reality*, HC 50–1 (The Stationery Office, 2009).

David Blockley, *The New Penguin Dictionary of Civil Engineering* (Penguin Books, 2005).

Chapter 1

House of Commons, Innovation, Universities, Science and Skills Committee, Fourth Report, *Engineering: Turning Ideas into Reality*, HC 50–1 (The Stationery Office, 2009).

Winston Churchill actually said 'We shape our buildings and afterwards our buildings shape us' in 1943 – about the re-building of the Houses of Parliament after World War II.

The Royal Academy of Engineering and the Engineering and Technology Board, *Public Attitudes to and Perceptions of Engineering and Engineers, a Study* (London, 2007).

Will Hutton, *Them and Us* (Little Brown, 2010).

Richard G. Lipsey, Kenneth Carlaw, and Clifford Bekar, *Economic Transformation* (Oxford University Press, 2005).

Toby Faber, *Stradivarius: Five Violins, One Cello and a Genius* (Pan Books, 2004)

Karl R. Popper, *Conjectures and Refutations* (Routledge and Kegan Paul, 1963).

Gillian Naylor, *William Morris By Himself* (Time Warner Books, 2004).

In a personal communication about the influence of technology on art, Michael Liversidge, an art historian at the University of Bristol, wrote, 'In his early years as a painter the palette of J. M. W. Turner was relatively confined, but as his career progressed a whole range of new pigments appeared that gave him a radically new chromatic range. So, for example, if one compares a painting such as *Calais Pier, An English Packet Arriving* of 1803 with *The Fighting Temeraire* of 1837, both in the National Gallery, London, the later picture has a brilliancy of colour that derives from new chemically derived pigments.'

An example of a piece of modern sculpture based on electronics is the 'Cloud' in Terminal 5, London Heathrow Airport. It is a five-metre-long shimmering disco ball created out of 4,638 electronic disks that flip from silver to black as controlled by a computer program.

In the UK, the Engineering Council is responsible for specifying engineering qualifications – see Engineering Council, *UK-Spec, UK Standard for Engineering Competence* (London, 2010).

In the UK, *engineering workers* are skilled qualified manual operators with a specific trade – bricklayers, carpenters, fitters, and many more. *Technician engineers* use known techniques to solve practical engineering problems safely – usually within a well-defined area of work. *Incorporated engineers* also work largely within an existing technology but have wider responsibilities. *Chartered engineers* are leaders with the widest scope of work and independence of mind. In many countries, such as the USA, chartered engineers are called professional engineers and have to be legally registered to practise in a country, state, or province.

In these five ages of engineering, we must also include the nuclear forces of nuclear power generation and the latest research in fusion, nanotechnology, and quantum computing.

Chapter 2

Although work is force times distance, the distance has to be along the line of action of the force.

SI is the International System of units – the modern form of the metric system. It is now the most widely used system of measurement throughout the world. A Newton is the SI unit of force that would give a mass of 1 kilogram an acceleration of 1 metre per second per second.

Before writing was developed, the objects of knowledge were stories held in the minds of people and passed on to successive generations.

Vitruvius, *Ten Books on Architecture*, ed. Ingrid D. Rowland and Thomas Noble Howe (Cambridge University Press, 2001).

Karen Armstrong discusses the Axial Age in her book *The Great Transformation* (Atlantic Books, 2006).

Thomas L. Heath, *A Manual of Greek Mathematics* (Oxford University Press, 1931).

Bertrand Russell, *History of Western Philosophy* (Unwin Paperbacks, 1961).

Virtual means not physically existing. The ideas of Aristotle and Jordanus de Nemore led to concepts of virtual displacements and the methods of virtual work which are now widely used in structural engineering.

Lynn White, *Medieval Technology and Social Change* (Clarendon Press, 1962).

John Fitchen, *The Construction of Gothic Cathedrals* (Oxford University Press, 1961).

Paolo Galluzzi, Istituto e Museo di Storia della Scienza, Florence, Italy, *The Art of Invention: An Interview with Paolo Galluzzi, from Craftsman to Philosopher: The Development of Engineering as a Profession*. See http://w3.impa.br/~jair/einter3.html

Francis Bacon, *Novum Organum*, ed. George W. Kitchin (Oxford University Press, 1855).

Chapter 3

For developments in the science of engineering, see: Hans Straub, *A History of Civil Engineering*, tr. Erwin Rockwell (Leonard Hill, 1952); and Jacques Heyman, *The Science of Structural Engineering* (Imperial College Press, 1999). Major theoretical contributions were made by men such as Charles-Augustin de Coulomb (1736–1806), who laid the foundations for soil mechanics; and Claude-Louis Navier

(1785–1836), who developed the theory of elasticity. Major practitioners were John Smeaton (1724–92), who developed the use of cast iron and of concrete; James Brindley (1716–72), who pioneered canals; Thomas Telford (1757–1834), the first President of the Institution of Civil Engineers; George Stephenson (1781–1848) and son Robert (1803–59), who developed the railways; John A. Roebling (1806–69), who designed major bridges in the USA, including Brooklyn Bridge in New York; and Isambard Kingdom Brunel (1806–59).

Henry Dirks, Introduction and Commentary on *The Century of Inventions, written in 1655 by Edward Somerset, Marquis of Worcester* (1864). See http://www.history.rochester.edu/steam/dircks/

10.3 metres is the maximum difference in height between levels of water in a pipe at atmospheric pressure.

Donald Schön, *The Reflective Practitioner* (Basic Books, 1983).

Chapter 4

For details of how genes may affect our choice of partner, see M. Bicalho, J. da Silva, J. M. Magalhaes, and W. Silva, 'New evidences about MHC-based patterns of mate choice', European Society of Human Genetics, European Human Genetics Conference, May 2009, Vienna.

John G. Landels, *Engineering in the Ancient World* (Constable, 1997).

A Leyden jar is typically a glass jar with a metal foil coat (that can conduct electricity) inside and out. A rod through the mouth of the jar is connected to the inner foil by a wire or chain and the outer foil is grounded. The jar collects charge through the rod so that the inner and outer foils store equal but opposite charges thus creating an electrostatic field.

The inverse square law says that the force F between two charges q_1 and q_2 is proportional to the inverse of the square of the distance r between them, i.e. $F = k\, q_1\, q_2/r^2$ where k is a constant.

Arc lamps create light by an electric arc or spark between two electrodes. They can be contained in and named after a gas such as neon or argon.

Heinrich Hertz (1857–94) was a German physicist who was the first to demonstrate satisfactorily the existence of electromagnetic waves.

For some details about harvesting power from human body heat, see V. Leonov and R. J. M. Vullers, 'Wearable electronics self-powered by using human body heat: the state of the art and the perspective', *Renewable Sustainable Energy*, 1 (2009): 062701. Also T. Starner and J. A. Paradiso, 'Human generated power for microelectronics', in C. Piguet (ed.), *Low Power Electronic Design* (CRC Press, 2004).

Thomas Kuhn, *The Structure of Scientific Revolutions* (University of Chicago Press, 1996).

Chapter 5

For some details about femtocells, see S. Saunders, 'Femtocells – a quiet revolution in mobile communications', *Ingenia* (September 2010), Issue 44.

It might be helpful to rehearse just how we count in decimal numbers. You may recall that the meaning of a decimal number such as 5,362.5 is $(5 \times 10^3 + 3 \times 10^2 + 6 \times 10^1 + 2 \times 10^0 + 5 \times 10^{-1}) = 5,000 + 300 + 60 + 2 + 0.5 = 5,362.5$

Let us examine a binary number such as $+ 1.011 \times 2^{100}$. The first part is the sign (plus is 0, minus is 1). The second part is the significand or mantissa $= (1 \times 2^0)$ before the point, and then after the point we have $(0 \times 2^{-1}) + (1 \times 2^{-2}) + (1 \times 2^{-3})$. If we add these together, we get $1 + 0.25 + 0.125 = 1.375$ in decimal. The third part is the exponent, which is 2 to the power $(1 \times 2^2 + 0 \times 2^1 + 0 \times 2^0) = 4 + 0 + 0 = 4$ in decimal. The total value in decimal is therefore $1.375 \times 2^4 = 1.375 \times 16 = 22$. In a computer, there are also typically 3 sections but in a different order. If the number is represented using 32 bits, then the first bit represents the sign, the next 8 bits represent the exponent, and the final 23 bits express the significand. The 1 before the point and the point itself are not included – they are assumed. Thus, in our example, we have one bit of 0 for the sign, then 8 bits of 00000100 for the exponent and 23 bits of 0110000000000000000000 for the significand. When put together we get 0 00000100 0110000000000000000000. In most computers, the exponent is offset by 127 in decimal, so the number actually stored would be 0 10000011 0110000000000000000000

Danny Bluestein, Yared Alemu, Idit Avrahami, Morteza Gharib, Kris Dumont, John Ricotta, and Shmuel Einav, 'Influence of microcalcifications on vulnerable plaque mechanics using FSI modelling', *Journal of Biomechanics*, 41 (2008): 1111–1118.

Chapter 6

Duty of care is a legal obligation not to act negligently under the law of tort. It is an obligation to take reasonable care to avoid foreseeable harm to another person or property.

The phrase 'fit for purpose' may be interpreted by lawyers as implying a strict liability as used in the purchase of simple products. Strict liability is liability regardless of culpability and is inappropriate for the provision of a professional service because it makes people risk-averse and inhibits innovation.

Reliability or safety engineering is widely used in chemical, oil, nuclear, and aeronautical/space industries and required by law in the USA and EU for the manufacture of hazardous chemicals. For some of the issues, see David Blockley (ed.), *Engineering Safety* (McGraw Hill, 1992).

For an introduction to the science of chaos, see Leonard A. Smith, *Chaos: A Very Short Introduction* (Oxford University Press, 2007), and for a demonstration of importance of initial conditions, see http://www.math.hmc.edu/~jacobsen/demolab/doublependulum. html

A system is vulnerable and hence not robust when small damage can cause disproportionate consequences.

For a discussion of *logos* and *mythos*, see Karen Armstrong, *Evidence for Religious Faith*, chapter 8, 'Evidence', ed. A. Bell, J. Swenson-Wright, and K. Tybjerg (Cambridge University Press, 2008).

Arthur Koestler coined the term 'holon' in his book *The Ghost in the Machine* (Picador, 1967).

For a technical discussion of structural health monitoring, see Keith Worden, Charles R. Farrar, Graeme Manson, and Gyuhae Park, 'The fundamental axioms of structural health monitoring', *Proc. R. Soc. A* (8 June 2007), 463 (2082): 1639–64.

Charles Perrow, *Normal Accidents* (Basic Books, 1984).

James Reason, *Human Error* (Cambridge University Press, 1990).

Royal Academy Engineering, *Engineering a Low Carbon Built Environment* (Royal Academy Engineering, 2010).

John E. Ahern, *The Exergy Method of Energy Systems Analysis* (John Wiley, 1980).

Further reading

Chapter 1

Matthew Crawford, *The Case for Working with Your Hands* (Viking, 2009). Popular good read about what Crawford and I see as the false distinction between thinking and doing.

Henry Petroski, *To Engineer is Human* (Vintage Books, 1992). On the nature of engineering written by an engineer.

W. Brian Arthur, *The Nature of Technology* (Penguin Books, 2009). Written by an economist with an engineering degree. The message is very much 'in tune' with this book. Less emphasis on technical principles but has well-chosen examples.

Engineering Council, *UK-Spec, UK Standard for Engineering Competence* (Engineering Council, London, 2010). Describes how to become a qualified engineer in the UK.

Carl Mitcham, *Thinking through Technology* (University of Chicago Press, 1994). A rare volume on the philosophy of engineering – a more in-depth treatment.

Chapter 2

Brian Cotterell and Johan Kamminga, *Mechanics of Pre-Industrial Technology* (Cambridge University Press, 1990). An easy read but contains some basic maths.

W. Harry and G. Armytage, *A Social History of Engineering* (Westview Press, 1976). Very good, easy read.

Geoffrey E. R. Lloyd, *Early Greek Science* (Chatto & Windus, 1970). A classic text.

Alistair C. Crombie, *Augustine to Galileo I* (Penguin Books, 1952). A classic text.

Alistair C. Crombie, *Augustine to Galileo II* (Mercury Books, 1952). A classic text.

Chapter 3

J. E. Gordon, *Structures: Or Why Things don't Fall Down*, 2nd edn. (DaCapo Press, 2003). A well-written account for a general reader about the way structures stand up.

Rolls-Royce, *The Jet Engine Book* (Rolls-Royce, 2010). A very nicely produced general introduction to jet engines by the makers.

David Blockley, *Bridges* (Oxford University Press, 2010). The science and art of some of the world's most inspiring structures aimed at the general reader.

Christian Wolmar, *Fire and Steam: A New History of Railways in Britain* (Atlantic Books, 2008). A readable history of how the railways came to be.

R. H. Barnard and D. R. Philpott, *Aircraft Flight: A Description of the Physical Principles of Aircraft Flight* (Prentice Hall, 2009). A good description of the physical principles of aircraft flight without mathematics.

Chapter 4

Steven W. Blume, *Electric Power Systems Basics* (John Wiley, 2007). Specifically aimed at the non-technical reader.

David Bodanis, *Electric Universe: How Electricity Switched on the World* (Abacus, 2006). A history of electricity – winner of the Aventis Prize for science books.

Brian Bowers, *A History of Electric Light and Power* (Peter Peregrinus, 1982). Comprehensive account from the nature of electricity to motors and power supply.

Paul Israel, *Edison* (John Wiley, 1998). Long, readable account of the life and work of a prolific inventor.

Tapan K. Sarkar, Robert J. Mailloux, Arthur A. Oliner, Magdelena Salazar-Palma, and Dipak L. Sengupta, *History of Wireless* (Wiley Interscience, 2006). Large, detailed history ranging from Maxwell's field theory to Marconi's antenna.

Chapter 5

Robert Plotkin, *The Genie in the Machine* (Stanford Law Books, 2009). Computer-based inventing in law and business.

Nicholas Carr, *The Big Switch* (W. W. Norton, 2008). Describes how computing is turning into a utility delivered through the internet.

Henry Kressel and Thomas V. Lento, *Competing for the Future* (Cambridge University Press, 2007). The story of technical innovation from transistors to venture capital and the future of digital electronics.

Malcolm Plant, *Teach Yourself Electronics* (Teach Yourself, 2003). A very clear introduction for anyone wanting to know some detail.

W. A. Atherton, *From Compass to Computer: A History of Electrical and Electronics Engineering* (San Francisco Press, 1984). A good survey requiring little technical knowledge.

Chapter 6

Gerd Gigerenzer, *Reckoning with Risk: Learning to Live with Uncertainty* (Penguin Books, 2003). A good read about risk.

Amory B. Lovins and L. Hunter Lovins, *Brittle Power* (Brick House, 2001). First published in 1982, this books shows how our infrastructure is vulnerable.

Barry Turner and Nicholas F. Pidgeon, *Man-Made Disasters*, 2nd edn. (Butterworth-Heinemann, 1998). Describes how human and organizational issues can incubate to cause accidents.

Peter Senge, *The Fifth Discipline* (Century Business, 1990). Excellent introduction to systems thinking.

David Blockley and Patrick Godfrey, *Doing It Differently* (Thomas Telford, 2000). Advocates systems thinking to deliver better major projects.

Index

STATISTICS
A Very Short Introduction
David J. Hand

Modern statistics is very different from the dry and dusty discipline of the popular imagination. In its place is an exciting subject which uses deep theory and powerful software tools to shed light and enable understanding. And it sheds this light on all aspects of our lives, enabling astronomers to explore the origins of the universe, archaeologists to investigate ancient civilisations, governments to understand how to benefit and improve society, and businesses to learn how best to provide goods and services. Aimed at readers with no prior mathematical knowledge, this *Very Short Introduction* explores and explains how statistics work, and how we can decipher them.

www.oup.com/vsi

The Roof Garden Commission

The Theater of Disappearance

Beatrice Galilee
Adrián Villar Rojas

The Metropolitan Museum of Art, New York

Distributed by Yale University Press
New Haven and London

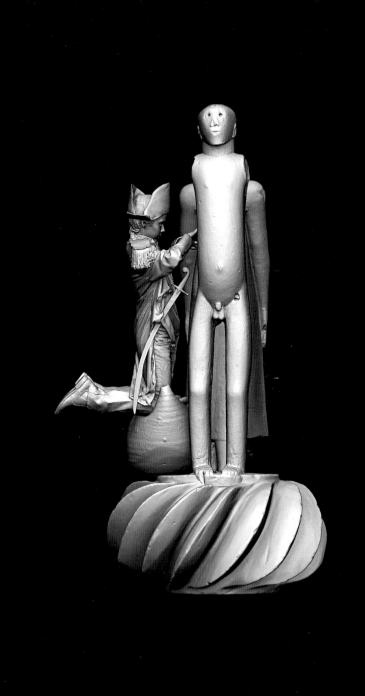

Sponsor's Statement

Bloomberg Philanthropies is proud to partner with The Metropolitan Museum of Art in support of "The Roof Garden Commission: Adrián Villar Rojas, *The Theater of Disappearance*." Villar Rojas's immersive, site-specific installation presents a unique experience to view objects from The Met's galleries, fused into imaginative new works of art and set against New York City's supreme skyline. We've been sponsoring exhibitions on the roof of the Museum for more than a decade, and Villar Rojas's work continues in the tradition of previous artists' by offering visitors an unparalleled encounter with contemporary art.

Bloomberg Philanthropies works to ensure better, longer lives for the greatest number of people. We focus on five key areas for creating lasting change: arts, education, the environment, government innovation, and public health. The arts are a valuable way to engage citizens and strengthen communities. Through innovative partnerships and bold approaches, the Bloomberg Philanthropies Arts program helps increase access to culture using new technologies and empowers artists and cultural organizations to reach broader audiences.

Bloomberg
Philanthropies

Director's Foreword

The Met's Iris and B. Gerald Cantor Roof Garden sits atop galleries that document more than five thousand years of world history and which have long been an unparalleled source of inspiration for artists. To realize the 2017 Roof Garden Commission, Adrián Villar Rojas immersed himself in the Museum's history and collections and held conversations with individuals across the institution, from curators and objects conservators to imaging specialists. His installation *The Theater of Disappearance* engages with our collection in a new and dynamic way, providing a provocative examination of traditional presentations and historical narratives that encourages visitors to reinterpret art and human culture.

Initiated by Sheena Wagstaff, Leonard A. Lauder Chairman of Modern and Contemporary Art, and organized by Beatrice Galilee, Daniel Brodsky Associate Curator of Architecture and Design, Villar Rojas's installation is a distinguished addition to our series of contemporary projects commissioned especially for this beloved outdoor space. We are especially grateful to Bloomberg Philanthropies for their long-standing leadership support of these much-anticipated Roof Garden presentations. We also extend our thanks to Met Trustee Cynthia Hazen Polsky and her husband, Leon B. Polsky, for their renewed gift to this year's installation and to the Mary and Louis S. Myers Foundation Endowment Fund for making this catalogue possible.

Thomas P. Campbell
Director
The Metropolitan Museum of Art

Activating Sculptures

Beatrice Galilee

At The Metropolitan Museum of Art, where does one start a visit but in the hallowed halls of the ancient worlds of Rome, Greece, or Egypt? To walk these galleries is to be humbled by a display of antiquity that represents a long-enduring vision of the pinnacle of civilization and aesthetic achievement.

Like many encyclopedic museums, The Met built its collection in the late nineteenth century with copies of sculptural masterpieces it could not yet afford to acquire. Plaster casts were inexpensive and accessible, allowing national academies and museums around the world, particularly in the United States, to exhibit the same emblematic sculptures found in European collections as well as architectural elements from significant ancient sites. The enthusiasm for such replicas led to intense academic study of and the public's fascination with classical art and architecture, informing a proliferation of artistic production from poetry to painting and sculpture. The subsequent development of an art-historical narrative of classical antiquity persists today.

For the 2017 Roof Garden Commission, *The Theater of Disappearance,* Argentinian artist Adrián Villar Rojas (b. 1980) both engages and disrupts The Met by drawing on this narrative and the five thousand years of world history within the Museum's galleries to create an elaborate ahistorical, environmental installation. In a startling interpretation of The Met's collection, highly detailed replicas of Museum objects form the basis of sixteen time-defying sculptures that fuse furniture, art, and artifacts, variously activated by casts of human figures sleeping, cheering, daydreaming, or kissing. To realize the spatial installation on the roof, Villar Rojas immersed himself in the Museum's history and collections and held conversations with art historians, scientific researchers, objects conservators, and imaging specialists. Over the course of several months, fascinated by the history and daily mechanics of the institution, he photographed items of interest or scenes that connected to his meetings or that prompted moments of reflection on the project at hand. Walking through the Museum's underground corridors, he stopped to take snapshots of the many archival images lining the walls. This publication, a document of the creative process behind the installation, comprises a selection from the thousands of photographs Villar Rojas both captured and found during his research, accompanied by his commentary.

When considering the dynamics of the Roof Garden, Villar Rojas had to reconcile its competing functions as both a gallery space for an annual contemporary art exhibition and a space to view Manhattan's expansive skyline. To unite the social and commercial aspects of the roof with his work, Villar Rojas has integrated each element into the conceit of a fantastical event. Decadent banquet tables occupy an oversize black-and-white checkerboard floor, punctuated by his exceptional sculptures, and visitors become partygoers as they order drinks at the bar and wind their way around the tables to survey the city from within Central Park. The resulting diorama, in which the public plays various parts in the larger performance of a party, constitutes a radical reinterpretation of the museum space.

As in all great works of staging, every component of the installation has been precisely considered and designed. As one approaches the tables, it becomes clear that each is an amalgamation of objects from varying world cultures combined with human bodies, coated in dust and marked with striations of time. In these sculptures, Villar Rojas has compiled artifacts that, to him, signify a tabula rasa of human expression and, together, a record of human mentality. For example, a nineteenth-century tomb effigy of Elizabeth Boott Duveneck from the American Wing is populated with sculptures: a lion and an otter from ancient Egypt watch over her, while an African cross is placed on her breast. A sixteenth-century Mexican vase lies overturned at her feet. Nearby, the figure of a woman wrapped in a duffle coat is found sleeping on the tomb of a medieval knight. In one hand, she holds a Mesoamerican mask, and in the other she grasps the blade of an Ottoman sword attached to an elaborate once-golden nineteenth-century French hilt. In another sculpture, a young boy dressed as Napoleon peers through two mirroring figures, one a nineteenth-century funerary post from Sudan and the other a more idealized carving of an Egyptian vizier from 1700 B.C. The installation serves to subvert the narrative put forth by geographically divided curatorial departments with a new generation of sculptures, all of which are inspired by repeating patterns that respond to the fundamental human concerns—such as reproduction, the harvest, and the afterlife—that have remained constant over the past five thousand years. The charged proximities that come from this merging of art and time, with an explicit contemporary human presence, evoke a playful but disarming response.

Preoccupied with the presence and politics of art, Villar Rojas's work elicits a sense of visual disturbance of time and space. Known for his large-scale installations, he deconstructs the status quo of every site, whether a forest floor, a museum, or a small island, usually altering its environment to generate entirely new conditions for encountering his art. These mostly temporary and always holistic interventions are often impossible to capture in photography and are left better understood as a kind of theatrical or cinematic experience, leading viewers to doubt the stability of their perception. He frequently imagines a scenario after humans have left the planet, questioning what the world might look like to an outsider life form, one that would not recognize the traces of our existence. By extension, he asks what artifacts might signify without the context and meaning we ascribe to them: Will a time come when a Converse sneaker can be equated with a masterpiece by Michelangelo as relics of a distant human epoch? This theme takes on new significance in *The Theater of Disappearance*, as he uses the Museum, a repository of the history of human culture, to investigate this question.

The concept of simulation provides a framework for Villar Rojas's method of interpretation and his reenactment of objects. In the treatise *Simulacra and Simulation*, the philosopher Jean Baudrillard reflects on the relationship between authenticity and meaning as ascertained through the perpetuation of symbols and signs. He highlights the story of the Benedictine monastery Saint-Michel-de-Cuxa, half of which was moved from its original location in France to later form part of The Met Cloisters in the early twentieth century, while the other half remained in situ, a practice of taking and leaving that mirrors archaeological excavations of the time. Baudrillard suggests that this act of acquiring only part of a site for display renders both what was removed and what remains as synthetic. Moreover, he posits that if the cloister were repatriated and reinstalled at its original site, the unified structure would be even more artificial than when half was in New York, as it would simulate reality by circumvolution. Such instability in the authenticity of public displays resonates with Villar Rojas's simulations of activating art-historical objects in the form of copies on the roof, with the originals just a few floors below. Like Baudrillard, Villar Rojas questions if replicas can replace or augment original versions and whether that changes our understanding of reality and history.

Through his work, the artist further plays with how reality is conveyed in museum presentations. Some sculptures that were once designated as manifestations of deities, for example, have been decommissioned from such roles and now stand as static artifacts of allegory and metaphor, cultural symbols stationed in glass vitrines or cabinets. Though an indication of a work's aesthetic, utilitarian, or ceremonial significance is often included in its display, a physical sense of the object has necessarily been lost. No method of exhibition captures the insects that once crawled in and around an object, or the wine carried in a jug, or the life behind the hand once ornamented by a ring. The exhibition of objects in a museum setting could be better understood as theory or fiction than as representation. *The Theater of Disappearance* signals a new chapter in the formulation of stories surrounding works of art, allowing these former gods to play out a new fiction.

The sixteen sculptures on the roof are the product of intersecting events, acts, performances, and conversations that took place across various platforms and around the world. Close to one hundred objects from The Met's collection were chosen by the artist after discussions with curators and thorough research into the Museum's agencies and adjacencies. They were then captured through photogrammetry or laser scanning by the Museum's photo studio. From his office in Argentina, Villar Rojas selected from photographs the human models who would interact with the objects from the collection, including babies, teenagers, and a septuagenarian, as well as a cat. Each person was assigned a costume or posture and then posed in front of 130 cameras, which generated a file of the body in three dimensions. The digital files were then grafted together to create three-dimensional models of the sculptures. Static objects were enlarged and twisted, some serving as items to be held or worn, others as cradles for human figures. The montages were passed from phone to phone, through e-mail and file transfer, until they arrived at a fabrication studio in New Jersey, where they were milled or 3-D printed, assembled, coated, and painted. A form of poetry, the sculptures bear the marks of the chisels or hands that shaped them, though no such tools were used in their fabrication. In fact, their production speaks to a new era that some call the fourth industrial revolution, in which fabrication will be entirely automated, thereby displacing the need for human hands.

The Theater of Disappearance could be the setting for an entirely new world, or the last party on earth, hundreds of years from now, featuring table centerpieces that look as if they were the petrified remains of humans performing everyday activities. It is a grand fiction of liberated objects being held and loved and touched, absorbing and re-presenting the imagery of Museum objects unconstrained by time and demarcations of culture or sequence. In probing the Museum's role in framing historical truth, Villar Rojas aims to disrupt its traditional presentations to allow for the reactivation and a reinterpretation of art and human culture.

Gods
and Ghosts

Adrián Villar Rojas

The Met's photo studio represents a hidden support system of Museum practice. This is where objects in the collection are reproduced in two and three dimensions for global dissemination in the form of postcards, catalogue images, and memorabilia.

A group of books that outline new frontiers in the preservation and replication of ancient artifacts is sandwiched between a prototype bookend of a first-century B.C. Assyrian palace guard and a replica representing half of a Corinthian capital.

Ron Street, The Met's first head of advanced imaging, sits in his office, a microcosm that brings together thousands of years of human culture in a few square feet.

The tracks at Laetoli, a site in Tanzania where a group of footprints were preserved in hardened volcanic ash, provided evidence for hominins walking on two legs 3.7 million years ago. They were documented in three dimensions and cast by a team at The Met using photogrammetry.

A replica of George Washington's teeth

A procession of objects leaving the tomb of Meketre in 1920 conveys the realities of early twentieth-century archaeological practices.

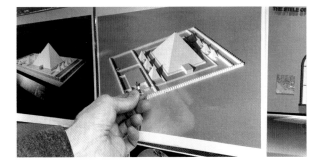

An Egyptian figurine is held up to observe a twenty-first-century re-creation of the pyramid complex of King Senwosret III (r. 1878–1840 B.C.) at Dahshur, Egypt, which was demolished to rubble during its own time.

hundred tombs in Western Thebes
between 1911 and 1936. The excavation
finds from the tombs are grouped here
by the various cemeteries designated by
the Expedition. A map of Western
Thebes situated near the north end of
the gallery indicates the locations of the
cemeteries represented, namely the 100,
500, 600, 700, and 800 Cemeteries. All
the objects come from the Metropolitan
excavations except for a group of
uninscribed pottery cones which were
found in Lord Carnarvon's excavations
between 1907 and 1913.

Material securely dated to Dynasty 11
and from the same cemeteries is on
display in Galleries 4a and 5. The objects
here—primarily of late Middle Kingdom
date—do not represent complete tomb
groups but rather a sample of the kinds
of material included in the burials of
middle-class citizens of Thebes. Many
are personal possessions used in daily life
that were placed in the tombs; others
are items which were strictly funerary or
ceremonial in nature, such as coffins,
models, and offering tables and trays.

From Thebes, Deir el Bahri, Sheikh
Abd el Qurna, Khokha, and Asasif,
MMA and Carnarvon excavations

Gift of the Earl of Carnarvon, 1913
Rogers Fund and Edward S. Harkness

A discursive text in the Egyptian galleries details The Met's expedition to Egypt
from 1911 to 1936.

The clearance of the tomb of Tutankhamun in the Valley of the Kings unfolded over a period of ten years.

Whether they manifest in objects unearthed from an Egyptian tomb or in representations on Assyrian reliefs, the preoccupations of ancient cultures remain consistent—the elements, the harvest, relationships with ancestors, fertility, death.

The brutal aftermath of the Battle of Gettysburg, in 1863, by Civil War photographer Alexander Gardner

Baskets and vases that were intended to carry the young pharaoh Tutankhamun to the afterlife are here rendered as objects to be labeled, boxed up, and shipped abroad.

Gardner staged this image of a dead Confederate soldier by moving his corpse, propping up his head, and setting a rifle at his side to emphasize the drama of the scene. Millions of minute gestures like this are constantly made in the field of history.

Existential questions and concerns take the form of offerings made to the deceased. Gold, basins, jars, and food were included in tombs to ensure that the deceased would protect the welfare of the group, the family, or the kingdom.

In 1870, shortly after the end of the Civil War, The Met was established in New York. Here, The Met stands in its early configuration in Central Park. Dozens of black-and-white photographs detailing moments of its history line the underground network of corridors used to move art through the building.

Among the marble monuments representing war and victory in the American Wing, *Mourning Victory* was commissioned after the Civil War as a memorial plinth for three brothers for a cemetery in Concord, Massachusetts. Four years after it was installed, the commissioner gave The Met funds to create a replica.

What if, during World War II, all cultural heritage accumulated by Europe had been lost under the reciprocal bombardment?

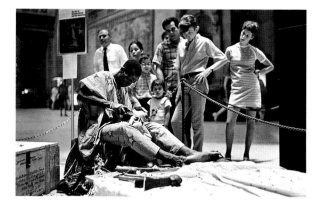

Could time and convention have turned the copies across US institutions into originals, enabling a resumption of the Old World?

In the autumn of 1939, museums in Paris prepared to close their doors in anticipation of war. The Hellenic marble sculpture Winged Victory of Samothrace (ca. 220–185 B.C.) was removed from the Louvre and hidden in a French château, along with the Venus de Milo and other prominent sculptures, until 1948. Perhaps in a mark of Allied solidarity for the catastrophic circumstances surrounding the sculpture's departure from public display, or simply in celebration of its glorious form, The Met placed a replica of Winged Victory at the top of the grand staircase in 1944.

Following the end of the Greek War of Independence from the Ottoman Empire in 1832, a treaty between Germany and Greece that allowed excavations in Olympia gave Germany the right to manufacture and disseminate casts of sculptures that were found, while the originals were to remain in Greece. The Greek Archaeological Service was established in 1833 to oversee all excavations, museums, and the country's archaeological heritage; it is still a powerful institution. The Met's founding constitution announced as its primary objective the formation of "a more or less complete collection of objects illustrative of the History of Art from the earliest beginnings to the present time." From 1883 through 1895, the Museum purchased two thousand casts of masterworks of Egyptian, Greek, Roman, ancient Near East, and Renaissance sculptures. In 1894, the Museum inaugurated the Hall of Casts, which functioned as the heart of the original building.

One can suggest that digging is a political gesture, or rather, that how deep we dig is a politically loaded action. Does what lies below our feet represent us? In Greece, they would say yes. In the former capital city of Constantinople, despite its common heritage of antiquity, the answer is often the opposite.

Metropolitan Museum of Art.

Tentative Lists of
Objects desirable for a Collection of Casts,
Sculptural and Architectural,
intended to illustrate the History of Plastic Art.

For private circulation
among those whose advice is sought in the preparation of final lists,
to enable them the more readily to make suggestions
to the Special Committee on Casts.

New-York:
Printed for the Committee,
June, 1891.

The Museum's only catalogue of its collection of 2,600 casts was published in 1908, shortly after its fourth president, John Pierpont Morgan, announced his disregard for casts as unfashionable and superfluous. Morgan disbanded the curatorship of casts, marking the moment in which the Museum was finally in a position to purchase original works of art. Nevertheless, for close to forty years, the installation of casts in the galleries remained unchanged.

The Museum's replacement of casts with original works can be read as a miniature history of the United States' growth. By 1948, the country entered prime economic and geopolitical conditions for its museums to acquire universal collections of original art.

Though the bulk of the cast collection sits in storage, some pieces are in their second phase of use. Here, two decorate the staff cafeteria.

A photo in an Aegean Airlines magazine shows President Obama walking among the monuments displayed in the Acropolis Museum.

The Department of Greek and Roman Art

What visitors experience in the Museum's galleries is the manifestation of power: the power of art and of America within the constructed, authoritative narrative of history.

One example of a conservator's work is this lion in the Department of Egyptian Art, which first entered the Museum as a plaster cast in the early twentieth century. The original was later purchased by The Met, and its fragmented nose was reconstructed. The lion—in particular, its fake snout—is the most-touched object at The Met. Tourists and school groups leave their human trace on a twentieth-century element of an ancient object.

In Ron Street's office, electricity gives agency to a figure of Christ. Because of this gesture, the wall is about more than just a compilation of past projects.

The Renaissance sculpture of Adam by Tullio Lombardo was damaged in a fall in 2002. After twelve years of research and digital scanning and modeling, the sculpture was restored.

Inside the Department of Objects Conservation laboratory, a conservator works on a medieval stained-glass landscape depicting farmers in the foreground and castles in the distance.

A small piece is missing among the castles at the edge, and he searches for images and research materials that would help him to faithfully reconstruct the lost fragment.

The conservator finds that the missing piece is not represented in the rectangular engraving from which the scene was copied.

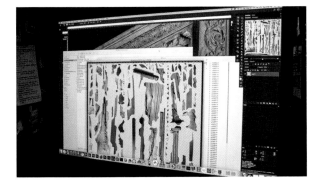

Should he invent, in the same way as the medieval painter of the stained glass did, the missing part of the castle? He knows that whatever he paints will instantaneously have the impact and authenticity of a thousand years of antiquity. Such is the legitimating agency of a museum, the most explicit manifestation of its authority.

Preservation is not a passive activity but a living force that extends past realities into the present and future. One of the most vital aspects of museum practice is the production of reality. How to display a work of art? Perfect, as it was minutes after being finished by its original creator? Or substantially traversed by time, deteriorated by the world?

What part of that deterioration will remain and become the reality of the object?

It is in these curatorial preferences that the ethical-political moment of conservation and preservation resides.

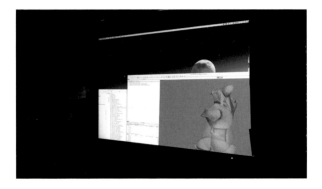

A three-dimensional rendering of the iconic Futurist sculpture *Unique Forms of Continuity in Space*, by Umberto Boccioni.

An offering: This Quanta System DNA laser technology used in the Department of Objects Conservation was a gift to the Museum from the Italian government in 2013 as part of the year of Italian culture in the United States.

The Museum's European period rooms branch out around the Petrie European Sculpture Court, offering a glimpse into the active lives of objects. The concept of the "butterfly effect," in which small causes can have spiraling consequences, can resonate in museum practice. The theory, popularized by the sci-fi author Ray Bradbury, is a real possibility in curatorial and conservation work, in which the museum is a studio where the text of human culture is constantly edited and reedited.

Information in the form of texts and maps accompanying works of art creates a sort of subtitle of a film whose original language remains unknown. Information can always be verified and contrasted, but as visitors travel through a museum, they surrender as children to its certainties: wanting to believe, implicitly trusting the voice of a mother or a father.

eated female

atite or chlorite, limestone
ntral Asia (Bactria-Margiana)
3rd–early 2nd millennium B.C.

Norbert Schimmel Trust, 1989
1.41a, b

e figures of seated females
luminous sheepskin robes
presented divinities. This
was also engraved on
d Elamite period
from southern Iran.

Monstrous male figure

Chlorite, calcite, gold, iron
Central Asia (Bactria-Margiana)
or eastern Iran
Late 3rd–early 2nd millennium B.C.

Purchase, 2009 Benefit Fund and Friends of
Inanna Gifts; Gift of Mr. and Mrs. Horiuchi, 2010
2010.166

A woman was seen wandering around The Met, murmuring to herself and seemingly full of angst and anxiety. At last, she put her hand in an empty square space of an Assyrian relief, as if there were some kind of magic involved in the act. Meanwhile, rumors circulate about a man who came into the Museum carrying his mother. This would have been fine except for the fact that she was inside a Folgers coffee can. He explained to the security staff that her dying wish was to be interred at The Met.

When security said no, he left his mother on a table and ran out of the Museum. The security office now has the can and is trying to find him. This is one among many tales of individual histories that mark the Museum.

For years, there have been whispers about the existence of a family of koi living in an empty elevator shaft inside the Museum, kept alive by an invisible cast of characters.

Under artificial light, the fish die and are replaced, as per the handwritten instructions taped to the heavily graffitied wall.

The American Wing could be regarded as the subconscious of the Museum: as in a dream, the rule of narrative formation seems to be metonymy and condensation. A series of meta-postcolonial interpretations manifests in an arrangement of sculptures in front of the Neoclassical facade of what once was a Wall Street bank. The rise of capitalism can be traced back to the conquest of the New World by the Spanish Empire. Gold and silver from Mexico to Bolivia arrived in Europe and formed a proto–international monetary system.

Before crossing the bank's threshold to enter a series of period rooms documenting US history, one encounters *Mexican Girl Dying*, possibly the most visceral sculpture in the Museum. A young woman lies wounded with a Christian cross beside her left hand. Is the uncolored marble a perpetuation of the white plaster casts of Greek and Roman sculptures?

In 1963, the *Mona Lisa* was exhibited at The Met after Jackie Kennedy proposed the installation to the French minister for cultural affairs. The painting was protected by bulletproof glass, and a guard was stationed on each side. In just three and a half weeks, one million visitors queued to see the famous work.

The facade of The Met, the collections within, and the presence of the *Mona Lisa* similarly drive hundreds of thousands of people to a state of procession: a pilgrimage to the temple of a secular God. Here, a human creation assumes the transcendent force of a divinity.

THE LIGHT OF HIS LIFE . . . Henri Bendel lights the cigaret of his wife after a dance. Plans for the Institute's new building were shown for the first time in the foyer of the hotel. A total of $700,000 has been raised toward the $1,000,000 goal.
Journal-American Photos by Sheldon Cotleur

The Met has, since its foundation, fused proximities of art, celebration, and festivity. The Costume Institute gala annually brings ancient Egypt into the contemporary economics of fund-raising and celebrity.

There is a simple agreement one enters into with museums: you can observe but you cannot touch.

Traditionally, a statue's placement in a public space is a strategic symbol of certain ideas, hierarchies, and values, whether political, religious, social, or economic. Most cities have public spaces where people congregate to express their concerns or their rejection of the status quo. These places are often chosen for their material representation of the establishment. In front of a courthouse, members of the Black Panther Party physically activate a Neoclassical-style sculpture.

The Theater of
Disappearance,
2017

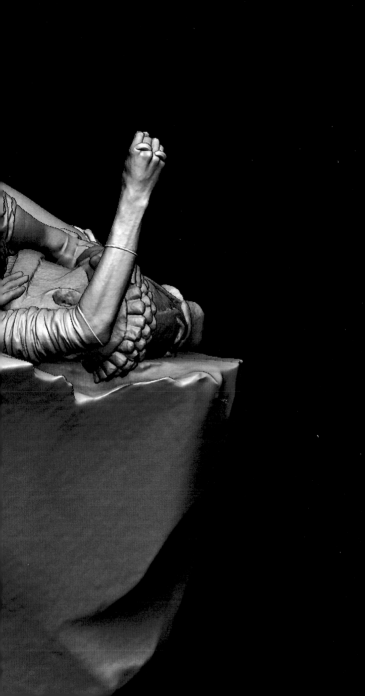

Selected Exhibition History

Solo Exhibitions

2004
"Incendio," Ruth Benzacar Gallería de Arte, Buenos Aires

2005
"Un mar," Alianza Francesa, Buenos Aires

2006
"Estas son las probabilidades de que te pase algo (nerd)," Galería del Poste, Centro Cultural Ricardo Rojas, Buenos Aires

2007
"Diario íntimo 3D," Centro Cultural Borges, Buenos Aires

"15.000 años nuevos," Belleza y Felicidad, Buenos Aires

2008
"Lo que el fuego me trajo," Ruth Benzacar Gallería de Arte, Buenos Aires

2010
"My Dead Grandfather," Akademie der Künste, Berlin

"Un beso infinito," Galería Casas Riegner, Bogotá

2011
"Poems for Earthlings," SAM Art Projects, Musée du Louvre, Paris

"La moral de los inmortales," Luisa Strina Gallery, São Paulo

2013
"The Work of the Ocean," Foundation de 11 Lijnen, Oudenburg, Belgium

"Today We Reboot the Planet," Serpentine Sackler Gallery, London

"Films Before Revolution," Zurich Art Prize exhibition, Museum Haus Konstruktiv, Zurich

2014
"Los teatros de Saturno," kurimanzutto, Mexico City

"The Evolution of God," High Line, New York

"Lo que el fuego me trajo: A Film by Adrián Villar Rojas," Galerie Marian Goodman, Paris

2015
"Fantasma," Moderna Museet, Stockholm

"Two Suns," Marian Goodman Gallery, New York

"Rinascimento," Fondazione Sandretto Re Rebaudengo, Turin

Group Exhibitions

2003
"Espacio Zona Emergente: Mariana y Adrián," Museo Municipal de Bellas Artes Juan B. Castagnino, Rosario, Argentina

II Salón Diario La Capital, Rosario, Argentina

"Currículum Cero 03," Ruth Benzacar Gallería de Arte, Buenos Aires

LVII Salón Nacional de Rosario, Argentina

2004
"Si a vos te pasa algo yo me mato," Subsecretaría de Cultura de la Universidad Nacional de Rosario, Argentina

XII Festival de la Luz, Santa Fe, Argentina

"Contemporary Argentinian Art," Melting Point Gallery, San Francisco

"Imágenes del Placer. Sabores del Vino," Patio de la Madera, Rosario, Argentina

LVIII Salón Nacional de Rosario, Argentina

2005
Bienal Nacional de Arte de Bahía Blanca, Argentina

"Sobre el amor (en el arte contemporáneo)," Centro Cultural Borges–Alberto Sendrós Gallery, Buenos Aires

"Cultura Pasajera," Pasaje Pam, Rosario, Argentina

"El dibujo es capital social: El club del dibujo," Centro Cultural Borges, Buenos Aires

LIX Salón Nacional de Rosario, Argentina

2006
"Amores posibles: 13 rosarinos / 7 ensayos," Zavaleta Lab Gallery, Buenos Aires

"Hola Data Entry, Hola Espectador," Centro Cultural General San Martín, Buenos Aires

Segunda Semana del Arte Rosario, Argentina

2007
"10 invitaciones premeditadas y consentidas," Museo Diario La Capital, Rosario, Argentina

"Cultura y Media," Centro Cultural General San Martín (Espacio Living), Buenos Aires

"Open Studio," El Basilisco Residence, Buenos Aires

2008
"25 X el Che: 80 años del nacimiento de Ernesto 'Che' Guevara," Museo Diario La Capital, Rosario, Argentina

"Nuevas incorporaciones," Museo de Arte Contemporáneo de Rosario; Museo Municipal de Bellas Artes Juan B. Castagnino, Rosario, Argentina

"Traducción (dibujos). Si la traducción fuera labor de humanos," Fondo Nacional de las Artes, Buenos Aires

"El concierto del año," Mite Galería de Buenos Aires

Further Reading

Adrián Villar Rojas: Poems for Earthlings. With texts by Rodrigo Alonso and Juan Valentini. Exh. cat., Jardin des Tuileries, Paris. Paris: Ed. du Regard, 2012.

Alonso, Rodrigo. "Argentina." In *ILLUMInations: 54*. Exh. cat., edited by Bice Curiger and Giovanni Carmine. Venice: La Fondazione, la Biennale di Venezia, 2011.

Boucher, Brian. "Adrián Villar Rojas Debuts Giant Sleeping David at Marian Goodman Gallery." Artnet news, September 10, 2015.

Burnett, D. Graham. "A Whaling in the Woods." *Parkett*, no. 93 (February 2014).

Chong, Doryun. "Grandeur Requires Violence, and Violence Makes Good Ruins." *Parkett*, no. 93 (February 2014).

Dorment, Richard. "Serpentine: Art Worthy of Zaha Hadid's New Building." *The Telegraph*, September 25, 2013.

Fiore, Alberto. "The Importance of Context and the Development of New Worlds." *Arte e crítica* 84 (Winter 2015–16).

O'Kelly, Emma. "Rotters' Club: Artist Adrián Villar Rojas and His Crew Hit Turin." Wallpaper online, December 7, 2015.

Preece, Robert. "Kneading the World from Scratch: A Conversation with Adrián Villar Rojas." *Sculpture*, June 2016.

Slenske, Michael. "The Nomad: On the Road with Adrián Villar Rojas and His Traveling Circus." *Modern Painters*, January 2016.

Smith, Roberta. "Review: Adrián Villar Rojas Explores Space in New Show." *The New York Times*, September 24, 2015.

Van Straaten, Laura. "Atop the Guggenheim, a Tiny, Secret Installation." *T Magazine*, September 8, 2015.

Villar Rojas, Adrián, and Sophie O'Brien, eds. *Adrián Villar Rojas: Today We Reboot the Planet*. Exh. cat., Serpentine Sackler Gallery, London, 2013.

Photography Credits

Acknowledgments

Adrián Villar Rojas thanks Mariana Telleria, adviser; Guillermina Borgognone and Germán Rodríguez Labarre, architectural design management; Joel Perez, drawing assistant; Hyejin Kim and Noelia Ferretti, production; Martín Paziencia, Matías Chianea, and Javier Manoli, sculptors; César Martins, engineering; and Malena Cocca, editorial design. For his performers, thank you to: Rishi Agnani, Banks Magid Baver, James Britton, Laurel Britton, Sandra Jackson-Dumont, the Frederick P. and Sandra P. Rose Chairman of Education, Beatrice Galilee, Ayesha Saveri Ghosh, Katyann Gonzalez, Carlie Guevara, Eric J. Henderson, Jaime Johnsen Krone, Lyle Krone, Terry Lovette, Helen Lykes, Jill Magid, Dario Fausto Stave Matias, Meryl Meltzer, Marine Pariente, John Rannou, Asad Raza, and Doug Williams.

We extend our warmest appreciation to Marian Goodman, Jessie Washburne Harris, and all at Marian Goodman Gallery, New York, London, and Paris, and to Mónica Manzutto and José Kuri of kurimanzutto, Mexico City.

At The Met, I extend my gratitude to Sheena Wagstaff, Leonard A. Lauder Chairman, Department of Modern and Contemporary Art, for her leadership; Thomas P. Campbell, Director, for his vision and direction; and Daniel Brodsky, Chairman of the Board of Trustees, for his unwavering support. Thank you to the individuals who lent their expertise to make this project possible: Alisa LaGamma, Ceil and Michael E. Pulitzer Curator in Charge, Department of the Arts of Africa, Oceania, and the Americas; Kim Benzel; Sylvia Yount, Lawrence A. Fleischman Curator in Charge, The American Wing; Diana Craig Patch, Lila Acheson Wallace Curator in Charge, Department of Egyptian Art; Maxwell K. Hearn, Douglas Dillon Chairman, Department of Asian Art; Luke Syson, Iris and B. Gerald Cantor Chairman, Department of European Sculpture and Decorative Arts; Joan R. Mertens; Pierre Terjanian, Arthur Ochs Sulzberger Curator in Charge, Department of Arms and Armor; C. Griffith Mann, Michel David-Weill Curator in Charge, Department of Medieval Art and The Cloisters; Niv Allon; Joanne Pillsbury, Andrall E. Pearson Curator, Department of the Arts of Africa, Oceania, and the Americas; Maia Kerr Jessop Nuku, Evelyn A. J. Hall and John A. Friede Associate Curator for Oceanic Art, Department of the Arts of Africa, Oceania, and the Americas; James A. Doyle; John Carpenter, Mary Griggs Burke Curator of Japanese Art, Department of Asian Art; and Pari Stave. Thank you to Barbara J. Bridgers, Ronald Street, Joseph Coscia, Jr., Oi-Cheong Lee, Hyla Skopitz, William Scott Geffert, Dahee Han, Christopher Heins, and Wilson Santiago in the Imaging Department; Quincy Houghton and Katy Uravitch in the Office of Exhibitions; Tom Scally and Taylor Miller in the Buildings Department; Amy Desmond Lamberti and Cristina Del Valle in the General Counsel; Clyde B. Jones III, Amy O'Reilly Rizzi, and Marilyn B. Hernandez in Development; Christopher Russell, York Ast, Natalia Granda, and Robert Burbano from Restaurant Associates; Emile Molin, Anna Rieger, Jaime Johnsen Krone, Brian Oliver Butterfield, and Bika Rebak in the Design Department. Thank you to Jon Lash and John Rannou at Digital Atelier, New Jersey; Doug Williams, Jordan Williams, and Graeme Williams at Captured Dimensions in Texas; and Rick Jordan and Carlo Masci. Above all, I am grateful to Adrián Villar Rojas.

We dedicate *The Theater of Disappearance* to the memory of Ronald Street, the Aleph of this project.

Beatrice Galilee, Daniel Brodsky Associate Curator of Architecture and Design, Department of Modern and Contemporary Art, and Adrián Villar Rojas

This catalogue is published in conjunction with "The Roof Garden Commission: Adrián Villar Rojas, *The Theater of Disappearance*," on view at The Metropolitan Museum of Art, New York, from April 14 through October 29, 2017.

The exhibition is supported by

Bloomberg Philanthropies

Additional support is provided by Cynthia Hazen Polsky and Leon B. Polsky.

The catalogue is made possible by the Mary and Louis S. Myers Foundation Endowment Fund.

Published by The Metropolitan Museum of Art, New York

Mark Polizzotti, Publisher and Editor in Chief
Gwen Roginsky, Associate Publisher and General Manager of Publications
Peter Antony, Chief Production Manager
Michael Sittenfeld, Senior Managing Editor

Edited by Elizabeth Gordon
Designed by Gina Rossi
Production by Lauren Knighton
Image acquisitions and permissions by Elizabeth De Mase

All artworks by Adrián Villar Rojas © Adrián Villar Rojas.

Additional photography credits appear on page 62.

Typeset in Galaxie Polaris and Chronicle
Printed on Opus 100gsm
Separations by Professional Graphics, Inc., Rockford, Illinois
Printed and bound by Puritan Capital, Hollis, New Hampshire

Jacket, cover, page 1, and frontispiece: © Adrián Villar Rojas

The Metropolitan Museum of Art endeavors to respect copyright in a manner consistent with its nonprofit educational mission. If you believe any material has been included in this publication improperly, please contact the Publications and Editorial Department.

The Metropolitan Museum of Art
1000 Fifth Avenue
New York, New York 10028
metmuseum.org

Distributed by
Yale University Press,
New Haven and London
yalebooks.com/art
yalebooks.co.uk

Cataloging-in-Publication Data is available from the Library of Congress.
ISBN 978-1-58839-621-1